Our Bodies: The Unexplored Link To God

Our Bodies: The Unexplored Link To God

◆

The Structure, Coloring and Functioning of the Human Body

J. H. Hacsi

Writers Club Press
San Jose New York Lincoln Shanghai

Our Bodies: The Unexplored Link To God
The Structure, Coloring and Functioning of the Human Body

All Rights Reserved © 2002 by Jacqueline H. Hacsi

No part of this book may be reproduced or transmitted in any form or by any means, graphic, electronic, or mechanical, including photocopying, recording, taping, or by any information storage retrieval system, without the permission in writing from the publisher.

Writers Club Press
an imprint of iUniverse, Inc.

For information address:
iUniverse, Inc.
5220 S. 16th St., Suite 200
Lincoln, NE 68512
www.iuniverse.com

ISBN: 0-595-24807-1

Printed in the United States of America

For Mom,
and Anne and Pat,
and Louis

No man is free who is not master of himself.
—Epictetus

If the mind, that rules the body, ever so far forgets itself as to trample on its slave, the slave is never generous enough to forgive the injury, but will rise and smite the oppressor.
—Henry Wadsworth Longfellow

Contents

Introduction xvii

Chapter 1 The Human Body 1
Did Some One or Some Force Have Something in Mind?

Chapter 2 Exceptional Organs 13
The Surplus Kidney and the Missing Penis/Uterus

Chapter 3 The Human Body…Is It Color Coded? 25
Can We Follow the Yellow Sun Lit Body to the Wizard of Us?

Chapter 4 Skin Color 37
There's More Than One Way to Become an Upright Biped

Chapter 5 The Laws of Heredity 47
When It Comes to Our Genes, Are We Like Peas in Their Pods?

Chapter 6 Heredity Versus Environment 59
Nature or Nurture: Which Force Forms Us?

Chapter 7 Envisionment 69
If Junior is "The Spitting Image" of Dad. Is He Sincere—or Sincerely Repressed?

Chapter 8 The Demon Theory of Disease 79
Modern Medicine: The Same Old Black Magic That We Know So Well?

Chapter 9 Failure and Success: Colds and Smallpox 91
Lose Some, Win Some…Is There Any Way to Better The Odds?

Chapter 10 Smallpox: A Closer Look 101
Did Dr. Jenner Have Help in Conquering the Pox?

Chapter 11   Yellow Fever 111
The Color Disease: Lemon Yellow Skin with Black Vomit

Chapter 12   Health, Good and Otherwise 121
If We Prefer the Good, Why Do We So Often Succumb to the Bad?

Chapter 13   Vision Problems 131
See Spot Run. See Dick and Jane Squinting to See Spot Run

Chapter 14   Arthritis 143
The Thigh Bone Is Connected To The Hip Bone, The Hip Bone Is Connected To The Backbone…

Chapter 15   Arthritis: Ready For Its Close Up?. 153
Dam Bones, Dam Bones, Dam Dry Bones

Chapter 16   Cancer, The Modern Plague 163
One?…One Hundred?…Two Hundred?…Three Hundred?… How Many Plagues Are We Up Against?

Chapter 17   Treatment: Conventional and Otherwise 173
If At First You Don't Succeed, Try…Try…Try…Try…Try…Try Again

Chapter 18   Cancer: The Mutant Cell 181
Under Too Rigid a Tyranny, Do Our Cells Rebel and Start the Revolution Without Us?

Chapter 19   Body-Squeak…Body-Speak 195
If Our Bodies Ran on Wheels, Would We Heed the Squeaking-Squealing?

Chapter 20   Verification. 207
To Test or Not to Test, Here Are Some Questions

Acknowledgments

My thanks to the following authors, associations, periodicals and publishers for the information they make available in print and on their web sites:

American Academy of Ophthalmology

American Cancer Society

American Psychological Association, publisher of *Psychological Abstracts* and the *PsycInfoDatabase*.

Anita's Book of Days, **http://www.halcyon.com/anitar/journal/080298.html**

Arthritis Foundation

J. A. Baldwin, "Schizophrenia and Physical Diseases," *Psychological Medicine*, 1979 (Nov.), Vol. 9 (4), pgs. 611-618, as summarized in *Psychological Abstracts*, Vol. 65, No. 1, Jan. 1981, pg. 359, Abstract No. 3339–J. D. Cooper.

The Baltimore Sun

W. H. Bates, *The Cure of Imperfect Sight by Treatment Without Glasses*, Central Fixation Publishing Co., New York City: 1920. Reprinted as *The Bates Method for Better Eyesight Without Glasses*, Holt, Rinehart and Winston, New York: 1940

Thomas R. Blakeslee, *The Right Brain,* Anchor Press/Doubleday, Garden City, New York, 1980.

David Bohm & B. Hiley, "On the Intuitive Understanding of Nonlocality as Implied by Quantum Theory," *Foundations of Physics*, Vol. 5 (1975)

Fritof Capra, *The Tao of Physics*, Shambhala Publications, Inc., Boulder, CO. 1976

G. Chedd, "Genetic Gibberish in the Code of Life," *Science 81*, Vol. 2, No. 9, Nov. 1981

Geoffrey Cowley with Mary Hager in Washington, "A Clue to Cancer," *Newsweek*, October 23, 1995, Section: *Lifestyle: Medicine*

Paul de Kruif, *Microbe Hunters*, Harcourt, Brace and Co., Inc., New York, 1926

Phil Donahue, *Donahue, My Own Story*, Simon and Schuster, New York: 1979

F. Eberson, *Microbe's Challenge*, Ronald Press, 1963

Encyclopedia Americana, Grolier, Inc., Danbury, CT: 1998.

Encyclopaedia Britannica, Chicago, IL: 2002.

Carol Enron, *The Virus That Ate Cannibals*, Macmillan Publishing Co, Inc., New York, 1981

Arthur S. Freese, *Help For Your Arthritis and Rheumatism*, The New American Library, Inc., New York, NY: 1978

H. Geist, Ph.D., *The Psychological Aspects of Rheumatoid Arthritis*, Charles C. Thomas Publisher, Springfield, IL.

Philip Goldstein, *Genetics Is Easy*, Lantern Press, New York: 1967

Saul Green, Ph.D., Book Review of *The Cancer Industry* by Ralph W. Moss, Ph.D., **http://www.quackwatch.com/04 Consumer**

Georg Groddeck, *Das Buch vom Es (The Book of the It)*, Vintage Books, 1949, 1961

Lucien Israel, MD, *Conquering Cancer,* translated by Joan Pinkham, Random House, Inc.: 1978.

Journal of Personality and Social Psychology, 1979 (Apr), Vol. 37 (4), 591-601

Harry A. Knopper, M.D., "The Truth About Cancer," copyright 1988, **http://user.icx.net/"drherb/cancer.html**

John Langone, *Aids: The Facts,* Little, Brown and Co., Boston, MA.:1988, 1991

Robert E. Netterberg, MD, and Robert T. Taylor, *The Cancer Conspiracy,* Pinnacle Books, Inc. New York, NY: 1981.

Wilder Penfield, *The Mystery of the Brain,* Princeton University Press, Princeton, NJ: 1975

The Pocket Book of American Slang, complied and edited by Harold Wentworth and Stuart Berg Flexner, Thomas Y. Crowell Co.: 1960, 1967.

Random House Dictionary Revised Edition, Random House: 1975, 1983

Robert Reid, *Microbes and Men,* Saturday Review Press: E. P. Dutton and Co., Inc.: 1974

J. I. Rodale and Staff, *Encyclopedia of Common Diseases,* Rodale Books, Inc., Emmaus PA: 1962, 1973

W. A. Rossi, *The Sex Life of the Foot and Shoe,* Saturday Review Press: E. P. Dutton and Co., Inc.: 1976.

Maggie Scarf, "Images That Heal," *Psychology Today,* Sept. 1980.

Bernie Siegel, MD, *Love, Medicine and Miracles*, Harper & Row, New York: 1986

"About Dr. Simonton," **http://www.arcobem. com/simonton/simonton. html**

R. M. Singer, "The relationship between visual refractive errors and interests, values, personality, academic performance and reading," as reported in *Dissertation Abstracts International,* B-Sciences, Jul., Aug. 78, No. 8933, pg. 1019B

Susan Sontag, *Illness as Metaphor,* Vintage Books/Random House, New York: 1979

C. Stern, *Principles of Human Genetics,* W. H. Freeman and Co., San Francisco, CA.: 1973

Jasom Theodosakis, MD., MS, M.P.H., Brenda Adderly, M.H.A., and Barry Fox, Ph.D, *The Arthritis Cure,* St. Martin's Press, New York: 1977

"Trophoblastic Hormones and Cancer: A Breakthrough in Treatment?," Session 205: June 13, 1998, **http://www.cmbm.org/ conferences/ccc98/transcripts/205.html**

University Microfilms International, Publishers of *Dissertation Abstracts International*

Craig Weatherby & Leonid Gordin, MD, *The Arthritis Bible,* Healing Arts Press, Rochester, VT: 1999

L. A. White, "Erotica and Aggression: The Influence of Sexual Arousal, Positive Affect, and Negative Affect on Aggressive Arousal, Positive Affect, and Negative Affect on Aggressive Behavior," *Journal of Personality and Social Psychology,* 1979 (Apr) Vol. 37 (4), 591-

601, as summarized in *Psychological Abstracts,* Vol. 64, No. 4, Oct. 1980, pg. 871, Abstract No. 8021.

Greer Williams, *Virus Hunters,* Alfred A. Knopf, Inc., 1959

Tom Wolfe, *Hooking Up,* Farrar, Straus and Giroux, New York, NY: 2000

James V. Writer, "Did the mosquito do it?", *American History, Jan/Feb97*

Yahoo Health Page, Rheumatoid Arthritis, Overview: Treatment. Copyright (c) 2001

F. A. Young, R. M. Singer and D. Foster, "Psychological Differentiation of Male Myopes and Nonmyopes," *American Journal of Optometry and Physiological Optics,* 1975–Oct., Vol. 52(10), pgs. 679-686, as summarized in *Psychological Abstracts,* Vol. 55, No. 1., Jan. 1976, pg. 755, Abstract No. 7478

C. A Zeiger, "A Psychological Approach To The Improvement of Myopia," as reported in *Dissertation Abstract International,* B-Sciences, June 77, Nos. 11-12, pg. 6359A

Introduction

*I am a little world made cunningly
Of elements, and an angelic sprite.*
—— John Donne

You who are reading these words almost certainly inhabit a body. Chances are that the body is human. Have you ever tried to analyze the physical form in which you find yourself living?

Most of us are so accustomed to the bodies in which we are tenants, are so used to being surrounded by other similar bodies, we pay no attention to the architecture of our forms. We may feel heartsick over our body's perceived shortcomings, its failure to live up to some culturally approved ideal, or we may be one of the small minority who happily flaunt a generous endowment in some area, but that's as far as it goes. Despite private moaning or public gloating, as long as our bodies possess the standard equipment and this equipment functions reasonably well, we enjoy the physical pleasures the body affords us without much thought. But consider this.

We are here. Somehow we arrived and are in possession of the particular bodies we have. Science claims that life on earth came into being by chance and, once established, evolved in response to environmental pressures. But what if science is wrong? What if it is zeroing in so intently on the specifics that it is blind to the overall picture?

At this point in our evolution, no one can prove that way back in the beginning we weren't lovingly planned for. When a manufacturer here on earth sets up a plant to manufacture washing machines, refrigerators or computers, and parts are ordered and laborers hired, the plant will almost certainly turn out the product it was set up to make.

Surely the creator of the universe had/has at least as much savvy as an earthbound manufacturer and could have planned the creation in all its myriad detail as carefully as the manufacturer of computers plans for the product he makes.

The fact that we might have been planned for does not, of course, prove that we were. But it seems more likely than the current creation myth put before us by science, which completely defies common sense.

No rational power would create a huge, spectacular universe and then deep-six it, letting it develop haphazardly, randomly, through accident, which is the current scientific scenario. Or if the argument is that there was no creator, that there was only nothing out of which our amazing universe somehow evolved, such a scenario, no matter now mathematically "sound" its backers may someday claim it is, is not only counter-intuitive but seems a little nutty besides. It is somewhat more rational to suppose that there *was* a creator and that this creator cared about his creation than it is to believe the opposite.

If this creative being or force knowingly set in motion the process through which we were created, then surely one of the most pertinent questions we can ever ask is: In creating the human body, did that being or force have something specific in mind?

> *Were our bodies carefully fashioned according to some as yet little understood blueprint to achieve some specific purpose, possibly in order to teach us something or to communicate with us?*

If our astronauts landing on the moon in July 1969 had found there a strange machine, the purpose of which they could not immediately comprehend, they would surely have studied it minutely, with great care, part by part as well as the interrelations between the various parts, in an attempt to determine for what purpose it had been constructed.

If we wonder why we are here, wonder about the purpose…or lack of purpose…of human life on earth, why not subject our bodies to the same scrutiny with which we would study some alien machine to see if by any chance some clues as to our purpose have been built into us?

Why is the human body formed as it is? Why do we have two arms, two legs, two eyes, two ears, two testicles or ovaries, but only one nose, one stomach, one liver, one brain and one heart?

Our bodies are cross-wired except for our nose, which is wired straight. Why?

Why does the human body come in the colors it comes in? Why are there no green or purple-skinned people? Why no orange or red-eyed people? Could the human body be color-coded?

Throughout history mankind has suffered recurrent plagues. As our laws and behavior change, so do our plagues. We outgrow some and grow into others. What can we learn about ourselves from this historical record of ever-changing plagues?

Why do our bodies become ill? Primitive man believed that demons caused disease. Modern medicine agrees, blaming demons they call germs. Isn't it time now to outgrow the primitive notion that demons/germs invade our bodies to do us ill?

Surely these are worthwhile questions to pose and ponder.

Before offering speculative answers to all of the above questions, I want to set forth the following hypothesis:

HYPOTHESIS:

> *Everything is one. All life is one.*
> *The structure, coloring and functioning of the human body are meant to teach us...or remind us...of this interconnectedness of all things, of the underlying unity of everything, that oneness is the heart of...and the reality of...all life in the universe.*
> *The purpose of life is to grow into an ever deeper awareness of this underlying unity, and to grow into an ever more intense love until we feel only love and are unafraid.*

* * * * *

This book, in looking closely at our bodies, attempts to put a solid foundation under this hypothesis. As I tell you of the clues I have seen, you can decide for yourself whether you see them too.

1

The Human Body

❖

Did Some One or Some Force Have Something in Mind?

o o o o o o o o o o o o o o o o o o o o o o o o o o o o o o
I will praise thee: for I am fearfully and wonderfully made.

—*Psalm 139:14*
Bible, King James Version

Science has taught us a great deal about the *how* of our bodily operation but little about the *why* of our construction. This book deals with the *whys*.

Why is the human body formed as it is?

To answer this question by responding, "Because. Because that's the way man has developed. Because that's the way life is and we share most of our physical characteristics with all other animals, or all other vertebrates," is to beg the question.

It's as though we were to answer the query, "Why does the Chevrolet automobile have four wheels and two headlights?" with the response, "Because the Ford car does." At some point in time, back before the first automobile was ever built, there was a planning stage in which it was decided, possibly by two eager young people in a barn, to construct the automotive machine with not three or five wheels but four and not one or three headlights but two. These numbers caught on, and the descendents of that original car now speed across our land with the predetermined number of wheels and headlights.

Let's assume for now that man too, like the automobile, is the product of a creative intelligence, a manufactured product almost the rival of the motorcar in importance, and that sometime back in the pre-production, drawing board stage for the animal life due to evolve upon the planet earth, there was similar planning. During this planning it was decided that life would develop along certain lines, for certain specific reasons, to ensure that Homo sapiens would, in time, emerge in triumphant, ideal form.

What is this triumphant, ideal form as we see it today?

Our bodies are made of bones, marrow, tissues, fluids, glands, plus a few accessories, for the most part wrapped in skin. Each body has one brain and one heart. We have a nose with fittings to deliver the oxygen we need, and a complete digestive system to handle fluid and nourishment, also a complete system for elimination of waste products, plus one penis or vagina/uterus for each body to allow for survival of the species.

In addition, we have eyes, ears, arms with hands and fingers, legs with feet and toes, plus teeth, nails and a generous amount of hair. Put all of this together and if all goes well we have a fully equipped and properly functioning human body.

Does this fully equipped and properly functioning human body have built into it any message regarding the ultimate meaning of life?

Imagine that you are faced with a crazed gunman whose finger is about to squeeze the trigger of the loaded gun he aims at you. The bullet that is about to strike you is a magical one. Once it hits its target, it stops in its tracks, doing no further damage. The gunman generously offers to let you pick his target. What part of your body would you offer up for destruction?

How about your hair? This is something you can easily survive without, and besides the chances are it will grow back in. There is often such a surplus of this accessory that our markets are filled with products designed to help us rid ourselves of the excess. Whoever designed our bodies was truly generous with hair.

If this doesn't satisfy the gunman, how about your nails? Or your teeth? Here again we are very generously supplied with something we can live without. For the mature human body, thirty-two teeth is a standard set. Lions and tigers, who would have a difficult time surviving in the wild without their teeth, have to struggle along with thirty, two fewer than we have. It's the same when it comes to nails. We have twenty per body, which is a surplus of four over the claws that the ferocious big cats have. How generously we are endowed when it comes to these decorative non-essentials.

If the gunman refuses to settle for external accessories, you could try offering an internal one, possibly the spleen, a vascular, glandular organ that is an accessory organ to the circulatory system. Or you could offer a gland such as the pancreas. Or better yet, your appendix, which you'll never miss.

If the gunman refuses these offerings, what would be your next choice? How about a hand or a foot, or, under pressure, an arm or leg? You have two of each and could survive with one. In truth, you could survive if all of your limbs were shot off, also if you lost both eyes and both ears. Although life would be diminished, it wouldn't end.

But why were we set up for such diminishment? When it comes to our limbs and our sensory organs, why is our endowment so limited? If we are allowed extra accessories, a greater number of teeth than the big cats have, a surplus of nails over claws, more hair than we need or want, why not extras when it comes to limbs and sensory organs?

Why should the human form be limited to two eyes? Surely an extra set of eyes, or even one extra eye, strategically located on the back of the head, would be of enormous survival value. And why not a second set of ears, possibly in the pelvic area or imbedded in our knees, to aid our hearing? Is there some specific reason that the sensory organs duplicated in the human form are not supplied in triplicate or quadruplet?

And why only two legs? As any infant knows, it's a struggle to learn to balance on two wobbly extremities. Why was the prehensile tail, if there was one, eliminated rather than strengthened and allowed to evolve into a third "leg" to help support and balance us, a la the kangaroo?

Why such wretched stinginess?

COMMENTARY

With two legs we stride upon our lovely earth. Using our two arms and our two hands, we maintain the personal home we have, our bodies. Through two eyes we see. Through two ears we hear.

Our feet feel solid as they stride upon the earth that also feels solid. Neither is. The earth and all creatures on it are largely empty space. The chair we sit on, the table upon which we write, both feel solid. Neither is. As all of us know, our senses lie to us.

The extent to which they lie is only now being explored by science. The current claim is that the world we see, hear, taste, feel and smell is not even a good imitation of the real thing. Reality is, science claims, a rainbow-hued pattern of uncountable interwoven waves, and we humans can absorb through our senses only a tiny, tiny fraction of this vast spectrum of waves.

In his book *Critique of Pure Reason*, published in 1781, Immanuel Kant wrote that time and space...two absolutes within our minds...have no independent reality, that they are instead organizing systems we have devised. Currently scientists are catching up with Kant, acknowledging that time and space are now being looked upon as illusions. The "reality" is now called "space-time."

This indicates that in an extremely fundamental way, we dare not trust even our own minds.

Using the information supplied by our sense organs as interpreted by our minds, we perceive everything in terms of duality: Night and day, male and female, material and non-material, life and death. And the most profoundly felt duality of all: The self and the non-self.

What if all the mystical reports throughout the ages are accurate and all of these perceived dualities are lies?

HYPOTHESIS:

> *We experience life through dualities for a reason. The creative force was not simply being stingy. Our senses, supplied in duplicate, feed us inaccurate information to help us understand that the dualities we perceive are only superficial "truths" and not the underlying reality.*

* * * * *

Faced with our crazed gunman, unless we are suicidal we would never choose to be shot in the brain or the heart or in most of our other internal organs.

If we look closely we will note that when it comes to almost every organ essential to human survival we are given only one. No multiplicity here. Even duplication is rare. *Why?*

> *Why did the creative intelligence we are postulating decide, in the planning board stage, to become so cheap and miserly when it came to the truly essential components of the human body? Why only one heart?*

Two lungs are provided. Why not a spare heart, with one heart placed symmetrically on each side, so that if one heart falters and fails the other can pick up the extra load and life will continue? With a creative intelligence or life force so seemingly addicted to symmetry and duplication…two eyes symmetrically placed on either side of the face, two ears on either side of the head, two arms on either side of the upper trunk, one leg extending down on each side of the pelvic girdle…why is this scheme not carried through in the chest with one strong heart beating away on either side?

Some earthworms have five hearts. While we can't contend that we are as important as earthworms, those busy little "farmers" who enrich the soil, couldn't we have been given at least two? Was there some reason that we were restricted to one?

The heart is not the only essential organ supplied singly. Without brain function life ceases, yet we have only one brain. One nose with fittings, without which we cannot breathe or live, one stomach, one liver. This seems a perilous state of affairs, to make life dependent upon organs supplied singly. It should prompt us to take a closer look.

For the sake of argument, let's continue to assume that the human form is not the product of a typically screwed-up committee, with every member tossing in her/his own pet idea no matter how idiotic, with the result determined by a politically driven majority vote, also that it is not the product of blind chance, a force that didn't know what in hell or in heaven's name it was doing.

Assuming instead…an unproven and for now unprovable assumption…that the human form was instead lovingly blueprinted, in every particular, by a supremely intelligent being or force, then we owe it to ourselves to look more closely to see if we can find anything approaching method in this apparent madness.

Why only one each of our most essential organs?

Let's look at the brain, one of the organs upon which human life depends.

Although we have only one brain, the largest and most human part of it, that part which evolved the most recently and governs voluntary movement and mental activity…the cerebrum…has the appearance of a double organ consisting of two identical hemispheres joined together by nerve fibers.

Does this mean that, with our highly developed nervous systems, we actually have two brains rather than one?

The best answer for this is in terms of how the brain functions.

If one of the hemispheres of the brain is surgically removed early enough in life, the remaining hemisphere will so organize itself that all brain functions will be met and the individual will grow up without impairment. This ability of the brain to adapt itself is soon lost, however. Once normal organization has been achieved, the surgical removal of either hemisphere will result in impairment. Each hemisphere, while working cooperatively with the other, is a specialist, and it takes both hemispheres...cooperation between two specialists...to cover the full range of activities governed by the brain. An apparent duality is, in respect to the way it functions, an operational unity.

Let's look at the heart.

The heart too has the appearance of a double organ. An interior wall divides it in half, each half consisting of a small upper chamber and a large lower one. If each half performed the same function, we would in effect have two hearts, just as we have two lungs and two kidneys. But the two halves of the heart do not duplicate each other's work. Here again an apparent duality is, in terms of functioning, an operational unity.

Regardless of the deceptive physical appearance, we have only one brain, upon which our life depends.

Regardless of the deceptive physical appearance, we have only one heart, upon which our life depends.

HYPOTHESIS:

The duality we perceive all around us...in particular the duality we feel, the me as separate and distinct from the not me...is also a deception.

* * * *

Q. Why was our one heart placed toward the left rather than toward the right or at a tidy, non-prejudicial dead center?

A. *We must breathe or we die, but one lung can collapse and quit and life will continue. Without brain function we die, but one hemisphere of the cerebrum, the largest part of the brain, can be traumatized or surgically removed and though some impairment will result, life will go on.*

The heart, though in appearance something of a double organ, functions as a tightly organized unity. The body can not lose half of its heart and live. If any part of the heart goes, life goes with it.

In this sense, the heart...every chamber of which is vital...is the most essential organ we have.

That part of the brain that governs the heart would become, by extension, the most essential part of the brain. The brain is composed, partially, of two hemispheres. The left hemisphere, which governs the right side of our bodies, is the verbal brain. The right hemisphere, in charge of the left side of the body, rules over our intuitive and feeling states.

If verbalization is the most basic and essential of our mental abilities, then the left brain should oversee our most basic organ, the heart. This would place the heart toward the right.

If verbalization and intuitive feeling are of equal importance, the heart should be dead center and under the rule of both hemispheres of the brain.

If our feelings, our intuitions, our dreams, are the most basic, essential part of us, of being human, then the right brain should oversee our most essential organ, the heart, and the heart should be placed toward the left.

Heart placement would be an extremely easy way for our designer, if we have one, to clue us in as to what matters most.

* * * * *

The human body is so constructed that it must be constantly replenished with food, water and air. To avoid dying of its own accumulated poisons, it must also rid itself of waste products. For this intake and outgo, the body has been furnished with several essential organs without which it will not survive.

With the primary sensory organs, and also with many of the body's glands, our designer...still assuming that we had one...seemed to go along with the old cliche that to put all of any body's eggs in one basket was risky. But this cautious, protective approach was abandoned when it came to the essential digestive organs. There is only one mouth, one esophagus, one stomach, one liver, one small and one large intestine.

When it comes to waste disposal, which is also essential for survival, the same situation prevails. With the exception of the kidneys, this work too is accomplished by organs supplied singly, that part of the large intestine known as the bowel, with an exit route through the rectum and anus, the bladder, the urethra, the skin.

The following simple statements would seem to cover the facts:

The human body is dependent upon nourishment for survival, and nourishment comes through the one. The human body is also dependent upon the elimination of wastes—purification—and this purification comes through the one.

HYPOTHESIS:

Our bodily construction shows us that survival depends upon the one, that life depends upon the one.
There are two exceptions to this rule, the lungs and the kidneys. Life is dependent on both lung function and kidney function, yet we have two lungs and two kidneys.
These two exceptions—the lungs and the kidneys—break the rule of bodily construction because each contains an important lesson for us.

* * * * *

Let's begin with a discussion of the breathing apparatus.

The nose, jutting forward from mid-face, is divided by a partition into two nostrils. Each nostril performs the same function of drawing in and expelling breath. In this sense, we have two organs joined under a common casing. So why were the two organs not separated as the eyes and ears are separated, with one nostril under each eye?

With breath essential to life...*And the Lord God formed man of the dust of the ground, and breathed into his nostrils the breath of life; and man became a living soul*...to separate the two nostrils, as the eyes and ears are separated, would seem to be a safer arrangement. Also it might prove to be a more attractive one, once we got used to it.

With two separate noses, each containing one nostril, each nostril supplied with a trachea or windpipe to send breath down into a lung, there would be a lessened chance of having the breathing apparatus severely damaged and put out of commission by a single blow. If one nose/trachea were smashed and rendered inoperative, the second could take over, adjust its workload, and life could continue without impairment. If I can see the advantage of such an arrangement, why wasn't it thought of back in the drawing board stage of human life, still assuming there was such a stage?

Possibly it was—and just as quickly discarded. There may have been a reason behind this slaphappy disregard of an easy safety precaution, an important clue as to meaning built into our bodies.

While physical life depends on food, water and oxygen, the most compelling of these needs is for oxygen. The body can survive for weeks without food, for days without water, but without oxygen the brain is irreparably damaged within minutes and death occurs soon after. Without ceaseless breathing in and breathing out, without the never faltering functioning of the breathing apparatus, physical life ends very quickly.

The breath of life. The breath that *is* life. How do we process the breath that is life?

We take in breath through our one nose, processing it through two nostrils, then sending it down one passageway, the trachea, into the lungs, where it is processed through myriad air sacs.

ODDITY:

The hemispheres of the brain cross over for their spheres of influence, the left ruling the right side of the body, the right the left. The nose alone does not share in the cross-switch from left to right and right to left. *Why?*

If any organ or organs are going to be exempted from the crossover, why not those sensory organs through which we learn so very much, the eyes or the ears? Or why not the heart, which upon we are so dependent? Why the nose?

HYPOTHESIS:

Our breathing apparatus, built to supply the most vital of our needs, was carefully fashioned to give us a blueprint as to the meaning of life, the mystery of being, the fact that life is a transformation of the one into the many. To accentuate this message, the nose alone is not cross-wired. It alone has the straight of it.

We breathe in life...come to life...through the one, the nose. We immediately split this unity, through our two nostrils, into duality, the separation of the created from the creator, spirit moving into and inhabiting material form.

Following this there is a rejoining, a recovered unity, as the one breath of life is sent down through the one trachea into the lungs. During the first few months of an infant's life, we are told, there is no consciousness of differentiation between self and non-self. Until a sense of self is developed, which takes time, all is one.

In the lungs, the trachea divides into two main bronchi...the duality with which we experience life once we develop a sense of self...before dividing into smaller and smaller passages ending in millions of tiny, balloon-like air sacs. In this way we process the one breath into the many, the myriad forms we see all around us. The lungs are an exception to a valid general rule that life depends upon the one because they are that part of the structure created to carry the message as to what life is all about: The one converted, for our pleasure or instruction, into the many.

* * * * *

In the following chapter we will deal with the two kidneys, and also ask the question: Why does each body have only one penis or vagina/uterus? These organs are not essential to individual life so why doesn't every body have two of one or the other?

2

Exceptional Organs

◆

The Surplus Kidney and the Missing Penis/Uterus

O, while you live, tell truth and shame the devil!

——*Edward de Vere*
17th Earl of Oxford
(Shakespeare)

Is there something unique about the penis or vagina/uterus that might have caused our designer, if we had one, to give it special treatment?

The male and female bodies for the most part are similarly equipped but, as many like to point out, there are differences. For one thing, the two bodies have different glands, one assortment for her, a different assortment for him. These glands manufacture the hormones that bring about sexual differentiation. A second difference, as we all know, is in their sexual apparatus.

Survival of the male or female body is not dependent upon possession of a penis or vagina/uterus, but the survival of the species *is*. (Medicine can now create new life without relying on the old-fashioned delivery system…male ejaculation in the vagina…but their methods don't always work and probably couldn't be adopted around the world fast enough to save the species if we couldn't fall back on nature and nature's way.). In this sense, and in this sense only, the penis and the uterus are essential organs. As procreative organs, they reaffirm what our bodily construction seems to tell us, that oneness equals the essential.

COMMENTARY:

Had the penis been blueprinted into the male body primarily as an organ of pleasure, it would not have been supplied singly. Had it been intended primarily as a source of sensation to add interest and enjoyment to life, the male body would have been supplied with two penises, not one, just as the female body would have been supplied with two vaginas, which might easily have doubled everyone's pleasure. When one penis/vagina had been exhausted, the spare could take over. One orgasm after another would have been the rule for everyone, not the exception.

However, unhappily, the penis and the vagina/uterus were blueprinted into our bodies as organs of procreation, not recreation.

* * * * *

Even if this is true, and there is absolutely no way to prove it, so what? In this day and age, when we are finally struggling free of the Victorian straitjacket, finally learning how fully and happily to indulge ourselves sexually, that's the last thing we want to hear, or, hearing, care about.

With the penis and the vagina/uterus, there is not only the slight problem, dealt with above, as to its proper functioning. There is also the interesting aspect of its placement.

Most pre-pubescent children, when they first hear how babies are made, react with a grimace and an "*Ugh!*" The whole idea makes them want to puke. They know full well that *down there* between their legs is a nasty place. Something smelly and dirty comes out that is not to be touched or played with. After it comes out you have to clean yourself carefully and then wash your hands with soap. The liquid stuff that comes out is worrisome too. By the time you're ready for school, or even before, you get jeered at if, in the heat of play, you wet yourself. Or, even worse, poop in your pants

Children deal with anxiety the same way adults do, by joking about what makes them anxious. Bathroom humor is big with kids of all ages. From recent movie reviews I have read...I haven't seen the movies...jokes about natural bodily functions are hot stuff on the silver screen right now, the more gross-out the better.

Why were we constructed in this fashion, with the organs of sexual love next door neighbors to the organs of excretion? Is there some purpose in this? Surely the primary organ of love, the penis, could have been placed several inches higher, possibly around the region of the naval, and also the vagina, thus removing them from that dirty place *down there.* Wouldn't this have confused us far less and done away with much of the guilt all too often associated with "natural" sexual pleasures? With this placement, the penis could have been excused from its excretory function and the male body supplied with the same urethral outlet as the female.

Why not? Was there some compelling reason this wasn't done, some specific purpose in the placement we have?

COMMENTARY

To remain healthy, our bodies must throw off waste material. Daily, or more or less often, our bodies eject solid matter. This matter, properly formed, is phallic in shape. To maintain physical health, we must eject a never ending series of penile forms. We must daily, or twice daily, or almost daily, give up the "penis" within ourselves.

Oh, shit, is it possible we are being told to give up sex?

To make this unwelcome "lesson" even clearer, was the organ of love...the penis or vagina...located right next to the anus, through which the feces are expelled, thus establishing a very strong intimacy by association?

This association of penis with feces seems so obvious that only the adult consciousness, carefully blinding itself to what it does not want to see, could fail to notice it. Very young children, infants and toddlers, easily make the connection. Many a young boy fights furiously against being toilet trained, against being forced to give up, to surrender, to have swept away down the toilet, a precious part of his anatomy, the very part that makes him like his daddy. He may protest this daily castration with every ounce of his strength, to the despair of his parents, who, failing to see any similarity between his stool and his penis, have no idea why he resists so hard.

Further evidence of this close relationship between feces and the penis is offered by some female bodies. If a tampon...phallic-shaped...is inserted into the vagina, a bowel movement...also phallic in shape...will automatically eject it in many though apparently not all women. To eject the feces is automatically to "give up" forms inserted into the vagina, and we all know what organ is most commonly inserted.

> *But why, why, why would a loving creator build such a message into our bodies, if indeed it is there?*

Throughout much of recorded history, there has been a tendency to link sexual abstention...celibacy...with goodness or godliness. In many world religions...Hinduism, Buddhism, Taoism, Jainism, Roman Catholic and the Eastern Orthodox Church...celibacy has a long, established tradition. It is seen as a discipline through which an individual can advance in spiritual growth, or as virtuous proof of devotion to God.

In the Upanishads, a collection of philosophical/psychological treatises composed by Hindu mystics between the 8th and 6th Centuries

BCE, ascetic elimination of all personal desires is propounded as the way to achieve unity with supreme bliss.

In the 6th Century BCE, Vardhamana Mahavira, who lived as an ascetic, founded Jainism by organizing a celibate clergy and an order of nuns.

Gautama Buddha, in the 5th Century BCE, formulated "Five Moral Rules" for right living, the fifth of which was, "Let not one be unchaste." For Buddha the basic evil is *tanha,* a selfish desire, above all sexual desire.

In the *Tao Te Ching,* a small book setting forth the beliefs of Lao-tse, who founded Taoism in the 6th Century BCE, it is written that the individual should not "act from any personal motive." He should govern himself to become again "as a little child."

Jesus encouraged the celibate life for his disciples: "Let those accept who can."

In the First Epistle to the Corinthians, St. Paul wrote, "I would that all men were even as I myself," that is, celibate.

St. Augustine believed it would be "joy unspeakable" if men could somehow procreate without having to engage in sexual intercourse.

Even where the above named religions were not in vogue, the linking of celibacy with religious belief or spiritual progress was not unknown. The Greeks had their Orphic cult, in which the subjugation of the flesh was aimed for as a condition of release of the soul, and the Romans their highly honored Vestal Virgins, white-clad, white-veiled women who took vows of celibacy and who, if found guilty of having sexual relations, were beaten with rods and buried alive, for who but virgins were fit to tend to the city's sacred fire?

Purity...chastity...goodness...godliness. Throughout history the linkage exists.

If there is any validity to this linkage, to the depressing, unhappy notion that celibacy is good for us, or helps make us good, our bodies, if we were creatively and intelligently designed, should bear witness to it.

Unhappily, in the placement of the genital organs, and in the fact that each body has only one penis or uterus with vagina, our bodies seem to do so.

* * * * *

The ancient yogis looked upon the human body and realized that it could be rather neatly split apart, metaphorically speaking, into different spheres of influence.

That portion of the body beginning at the waist and extending down through the two legs to the feet can justifiably be considered the physical portion of the body. The major digestive, eliminative and reproductive organs are located here, and the legs/feet connect us with the physical earth.

That portion of the body from the waist to the shoulders can justifiably be considered the emotional portion of the body. The heart, universally considered to be the site of feeling, and the lungs, which process the "breath of being," are located here, also our arms/hands extend from this area, and it is with our arms/hands that we reach out to others, either to give or to take.

From the shoulders up is the mental, the head housing the brain.

A further symbolic division was made of the face. The lowest third, containing the mouth, which connects directly with the stomach/digestive system via the esophagus, was seen as the physical portion of the face. The mid-section, containing the nose, which connects directly through the trachea with the lungs, was seen as the emotional portion. The top third, containing the eyes, which connect directly through the optic nerve with the brain, was seen as the mental portion.

Keeping this helpful topography in mind, I'd like to ask a couple of questions and offer speculative answers:

Q. Why does the female body have a much more pronounced waist than the male body?

A. *If the two portions of the torso are drawing apart, if the emotional is becoming less based upon the physical, this should appear where the two parts converge, at the waist. This is the place to cut them off at the pass. The female body does this to a much more pronounced degree than the male body, indicating that women are ruled more by their hearts and less by their groins than men. Observation of most cultures would bear this out.*

Q. Why is the male body customarily far more hairy than the female body?

A. *Hair is quite intimately tied in with the sexual. At puberty hair begins to grow where it hasn't been before, in the genital area, under the arms, on the male chest. Those ruled more by their groins and less by their hearts would quite naturally grow more hair as a manifestation of where their ruling passions lie.*

* * * * *

Why two kidneys? When kidney function is essential to life, the fact that we have two kidneys blows away the entire speculative scheme (oneness equals the essential) if we can't come up with some plausible reason that we have a second kidney.

All the blood in the body flows through the kidneys every few minutes. Waste products of living cells are removed, plus any excess water and acid. This waste fluid...urine...leaves each kidney through a tube, the ureter, and empties into the bladder, a storage sac. It is then expelled from the body through another tube, the urethra.

In the male body, the urethra is sheathed by the penis, which not only functions as a semen-delivery system but also for the discharge of urine. A sphincter muscle at the base of the bladder, under both voluntary and involuntary control, tightens as penile erection occurs, shutting off the flow of urine to keep the two fluids, semen and urine, from intermixing. At any time the penis, enclosing its urethra, can be used to urinate or ejaculate, but it cannot do both simultaneously. A choice must be made.

This is a curious situation. With no other organ of the body is this type of decision forced upon us. We can choose to close our eyes or keep them open, but we do not have to decide whether we will use our eyes to see with or to hear with. The same holds true for our ears. Though we use our noses both to breathe and to smell, these are done concurrently. We don't have the power to decide that for the next couple of minutes we'll breathe, then after that we'll smell. We don't have to decide whether to have our hearts pump blood or switch to some other task, nor do we have any control over the function the brain will perform. Our digestive system does its own thing with no go ahead from us, as does most of the purifying apparatus. Only the penis, with its dual, unrelated functioning, forces a choice.

Fortunately for the continuation of human life, this choice between urination and ejaculation can be made moment by moment, with a constant switching back and forth. Males can choose to urinate at intervals all day, then ejaculate once or twice before falling asleep. Were this not the case, if the choice had to be made once and for all, irrevocably, at, say, twelve noon next Tuesday, life would soon dribble to an end, *regardless of the choice made.*

If the choice were for ejaculation, the male body would soon die of its own accumulated poisons. If the choice were for urination, the species would soon die out for lack of procreation.

Thankfully, the sphincter muscle...supplied singly to each male body...allows the human race to continue without forcing this choice. As long as this muscle continues to tighten and loosen, tighten and loosen, without ever getting permanently stuck in either the *on* or *off* position, we will survive.

Could we have two kidneys because the urinary system spells out for us that we make choice after choice as to how to live our lives?

During our habitation of the human bodies we find ourselves in, though we share membership in the same species, we all live…in a sense…on different levels. On one scale we can experience anything from robustly glowing good health to dismally weak, debilitating ill health. In like fashion, we can live anywhere between two extremes when it comes to our interaction with others.

The man who rapes and kills a child is attacking life from a different perspective than the man who, at great risk to his own life, will plunge into a burning building to save a child. The man or woman who smuggles into the country illegal drugs, or deals in them, lining up young people to sell to other young people, lives life on a somewhat different level from the man or woman who devotes his or her life to caring for and teaching handicapped children. In all of these cases, a life is being lived, but the lives do not necessarily equate. A man who rapes and kills is subtracting from the sum total of health and happiness in the world while the man who saves lives is adding to this total.

We can live a life that most people would see as a loving one, or one that is not so loving. The choice is ours, a choice we make over and over as long as we occupy our bodies. Could this be the reason that we have two kidneys?

COMMENTARY

The lower portion of the body represents the physical aspect of life and this is where, in the male body, the penis is found. The penis can be thought of as the organ of desire/lust…lust for sex, for money, for power. We can conquer this lust or be ruled by it. If we are ruled by it, if we choose lust for the physical over all other values, we are exercising the penis, possibly over-exercising it, choosing the penis' lust-driven function over its excretory one. This choice in effect renders the kidneys relatively inoperative.

If this is not a good choice, how could our designer, still assuming we have one, let us know that it isn't?

One way might be by furnishing us with an extra kidney. This alerts us to the fact we lead a dual existence.

In being...existing...we lead a physical life. In action...interaction with others...we lead what could be called a moral life.

The duality of our kidneys might be a way of letting us know that an active kidney promotes physical health. In addition, as it renders the penis' sexual function relatively inactive, it promotes moral well being.

* * * * *

Q. Why are our kidneys of different size and shape? The right kidney is a little lower and smaller than the left. The left kidney is higher, longer and narrower than the right. Why?

A. *Our moral life is ruled more by feelings than by words. The left kidney...the heart-side kidney, the kidney under the rule of the intuitive right brain...is a little larger and a little higher than its near twin for this reason: It has a slightly larger job on a somewhat more elevated plane. For this reason it is longer and narrower, as the road to heaven is said to be.*

SUMMARY

If we are not the products of blind chance but were instead lovingly designed, the designer might easily have built clues as to our meaning and purpose into our structure.

If we look carefully enough, with open minds and hearts, we can zero in on some of the clues. Our bodies clearly show that our lives depend upon the one, while we are also supplied with roadmaps...the lungs, the two kidneys, the sexual apparatus...to help us navigate the journey in material form that we call life.

*****

The human body has more than form to guide us. The form has color. Let's look to see if are any clues to be found in our coloring.

3

The Human Body... Is It Color Coded?

❖

Can We Follow the Yellow Sun Lit Body to the Wizard of Us?

○ ○
The heart should feed upon the truth, as insects on a leaf, till it be tinged with the color...

—*Samuel Taylor Coleridge*

We live in a world filled with color. Color not only delights our senses but is also an easy way to give instruction and direction. Dorothy was told to follow the yellow bricks to the Emerald City of Oz. In our own day, the flow of traffic is ruled by color. Hospitals, business establishments, recreation areas often have color charts embedded in their floors. Items that are purchased unassembled often use color tips to aid the new owners in their assemblage. This use of color permeates our lives.

If we, with our limited intelligence, have been bright enough to develop this use of color, why couldn't our creator, if we have one, have thought of this too? Possibly the artist who painted the created universe did not haphazardly splash color around but instead carefully color-charted the entire creation.

Let's assume for the sake of discussion that this did indeed happen, that colors were not thoughtlessly thrown at the universal canvas but were instead carefully thought out and assigned. The human body, an interesting part of the creation, a body that would in time take its place as undisputed kingpin of the earth, would have been included in this procedure and had its colors carefully assigned too.

With this assumption in mind, let's take a look at the coloring scheme in us, on us and around us, excluding from consideration, for the moment, that rarity, the albino, on whose body the normal color pigmentation is lacking.

Consider the eye. The pupil of the eye, an expanding and contracting opening in the iris through which light passes onto the retina, like the early Ford car comes in one color only, black. The iris, in contrast, comes in a variety of colors, yet from a very limited range. Why, among those with normal coloring, are there no healthy red irises, or orange, or pink?

The exterior wrapping of the human body, the skin, comes in a variety of colors also, yet again from a limited range. Why is there no race of blue-skinned people, or purple, or green?

Man does not exist in a vacuum. Human life exists against the backdrop of a big, beautiful, colorful world. If we make note of the permissible range of colors of various bodily organs, then relate this range to the appearance of color in the world around us, possibly we can detect the great designer's "color code" and learn to break it.

Of the three primary colors...red, yellow, blue...one dominates our physical world. The vast expanse of the sky, as seen from the surface of our earth during daylight hours, is blue. Close to three-quarters of the earth's surface is covered by water, and the oceans, seas, rivers and lakes of our planet home are predominately blue. This would seem to indicate that the great artistic designer who created our world, if there was one, was extremely fond of the color blue. So why aren't we...an important and unique creation...wrapped in a lovely blue skin?

Turquoise, sky blue, royal blue, violet, purple. Why are these not the races of man?

In contrast to the splashy, lavish use of blue in the physical world, when it comes to the human form, the appearance of blue is very scanty. In the absence of artificial dyes or genetic abnormality, and as long as we are alive and well, there are no blue skins, blue hair, blue teeth or blue nails. If our coloring is fair enough, we can glimpse blue in the veins under our skins. Apart from this, blue appears only infrequently in Homo sapiens, and then just a tiny circle of it in the iris of certain eyes.

Only two of the primary colors seem permissible for irises: blue and yellow. Except for albinos, where, in the absence of normal color pigmentation, blood vessels give the appearance of a pinkish hue or redness, eyes can be blue, blue-gray, gray, blue-green, green (blue/yellow), that darkened shade of yellow known as brown, hazel (a light golden brown) or black (all colors combined). The third primary color, red, is eliminated in all its infinite variety, except as a component of black. The iris of the human eye is never naturally, healthily red, nor any shade thereof. *Why?*

For the largest and most visible organ of the human body, the skin, there is a switch in the primary colors allowed. Red and yellow are in. Blue is out. *Why?*

Why are blue eyes allowed but not blue hair, blue nails, blue teeth or blue skin?

If we check out the role that the color blue plays in the universe around us, maybe we can stumble onto a speculative answer.

COMMENTARY

Man locates heaven up in the sky, in the vast calm blue. If this is a proper placement, then peace...happiness...heaven...are to be found in that which is blue.

Hair is sexy. If heaven is to be found in sex, hair should be blue, like the sky.

Nails and teeth are bodily weapons. If heaven is to be found in warfare, in assaulting others or defending oneself, then our nails and our teeth should be blue.

The skin is our bodily surface. If to stay on the surface...touching, being touched, going no deeper...will bring us to heaven, the skin should be blue.

If, however, heaven is to be found in none of the above but rather through an elevation of interest from the physical through the emotional to the mental, then the eyes...located on the mental level, connecting directly with the brain...would be the one bodily organ allowed to share the heavenly blue of the skies.

* * * * *

In the normally colored human body, red has been banished from the colors allowed in the human eye. Let's check for red in the world around us, and in ourselves, to see if we can find a possible reason for this.

All life on earth is slavishly dependent upon the sun. While the sun is yellow/white all day in our sky...when we can see it through the clouds, haze, rain, fog or smog...it is orange/red when it rises and red when it sets.

The fluid which courses through our bodies, bringing nourishment and growth, is red. When cut, we spurt red. Menstrual blood, signaling no new life has been conceived during that cycle, is red. The erect penis, engorged with blood, in the fair-skinned body flushes pinkish/

reddish. Lips, both facial and genital in the female, are pink or reddish, as are nipples.

For us, as with the sun, the beginning of life, sexual activity, is reddish-hued, and death, the end of it, if untimely and violent, is often red too.

This may answer our question as to why, in normally pigmented bodies, there are no red eyes.

COMMENTARY:

We can live our lives primarily on a physical level, paying attention to the demands of the penis, to its lust for sex, for money, for power over others. Living in this fashion, we can enlist the rest of our bodies, our hands, arms, sensory apparatus, lungs, heart and brains, into penile servitude, expressing this attention to the penis by being satyrs or rapists, thieves or muggers, con men, drug dealers or politicians.

If this doesn't do it for us, we can elevate our interest and live on an emotional level, waist to shoulders, listening to the yearnings of our hearts for a more intense life. Living in this fashion, we will put our body's energies at the service of the heart and its enjoyment of those delicious excitations and delightful palpitations that come with falling in love, with the quickening of the pulse and the aching wish to hold and be held, to love and be loved, endlessly, endlessly in love.

Or, if that doesn't quite do it for us, we can elevate our interest one more level, to the mental, and put our physical and emotional natures under the control of the mind, letting the mind direct us.

The eyes exist on the mental level, connect directly to the brain, and should be colored accordingly, clueing us in as to the profit or loss of the mental life.

Red is the color of violent death, the color that daily precedes the plunge of our world into darkness. In the direction of traffic, red signals STOP. All over the world, while green, an allowable color in the eye, means safety, red means danger.

If the elevation of interest from the physical through the emotional to the mental is dangerous, if it means a plunge into darkness, into death, then red would be an appropriate color for the iris of the eye.

On the other hand, if such an elevation of interest isn't dangerous but instead signals a rapprochement with the blue of the sky, the peace and calm of the heavens, or a linkage with the waters of life, then red should be ruled out as a color in the human eye.

* * * * *

What of green, that lovely color that *is* allowed in the eye?

As we look at our world, particularly during the marvelous, bursting days of spring, we can have little doubt that green is the color of renewal. When the yellow of our sun blends with the blue heaven of our skies, rebirth results. Yet green, very much like blue, can be found visibly in the human form only occasionally in certain irises.

The green of renewal, like the blue of heaven, is restricted to an infrequent appearance in the eye.

Yellow, in specks or in that darkened form we call brown, is also a permissible color in the eye. Is this a plausible coloring?

It is through our yellow sun that we see. Without light we would all be sightless. Yellow clearly is the color of seeing, of awareness, of knowing.

HYPOTHESIS:

> *The range of permissible eye coloring clues us in as to the merits or dangers of living primarily on the mental plane, of allowing our minds to rule us instead of our physical lusts or emotional yearnings.*
>
> *Red is not an allowable color as there is no danger in this approach to life.*
>
> *The colors that are allowed in the eyes tell us this:*

If those of us in human bodies would find peace (blue) or renewal (green) we must go to the mental level, to our eyes. We can also find awareness (yellow) there, in a darkened or speckled form, but no one of us housed in a human body sees well enough, clearly enough, to have pure yellow eyes. To paraphrase St. Paul, we can only see through a lens darkly, not with the full light of the life-giving sun.

* * * * *

Red means danger, everywhere in the world. It is the color of the spilling of blood and the color of the dying sun just before our world is plunged daily into darkness.

Apart from this, and the pinkish/reddish tint of our genitals, lips, nipples, and nails, where else do we see red?

When angry, we see red. We can be red-faced with shame, or red in the face and neck due to the angry red blood pounding through us. If anger has a color, that color is red.

Yet if we look at our bodies...at our genitals and those other bodily parts most apt to participate in sexual contacts...and assign a color to sex, that color would also have to be red.

If sex and anger share a color, as they do, is there any connection between the two?

Is anger a primary emotion in sex? Is sex an expression of anger?

American slang would seem to answer these questions in the affirmative. The common interjection *Fuck you!* is not meant as a loving suggestion. To give someone the finger is both sexually suggestive and an open expression of anger and contempt. *Go stick it...Stick it up yours...Up yours,* verbal equivalents of the gesture, do not express a friendly sexual interest. The world "attack" can mean assault in public or private war, with penile like-weapons or with the penis itself.

Studies with monkeys have shown that anger is an essential ingredient in sex. Another study proved that the same was true for rats. Rats were placed in luxurious surroundings, with ample space, an overabundance of food, all the amenities for the good rat life. Mating became less frequent, then ceased altogether. Without frustration to spark anger, there was no sex drive.

Observation of non-human animals in the wild has shown the close tie between non-sexual assaults and sexual activity. A young male caught in a vulnerable position by an older, stronger male where flight is not possible will drop into the passive female sexual posture, allowing the stronger male to mount him. The older male with make a few thrusting moves, then will dismount and amble away, his aggressive mood defused.

A study by L. A. White among male college students showed the same underlying connection between non-sexual assaults and sexual behavior. Subjects were deliberately angered, then given...they were led to believe...a chance to retaliate. Prior to the opportunity for retaliation, most were shown erotic material in one of four subgroups: With high or low positive erotic content, or high or low negative erotic content.

Those who watched positive erotic stimuli significantly *reduced* their retaliatory behavior compared to a control group, while those who were shown the negative material were slightly *more aggressive* than the control group.

Anger can be channeled into sexual arousal. Blockage of sexual arousal increases the level of anger.

SUMMARY:

When blood circulates through the body in its proper channels, it carries with it health and growth. When it bursts or leaks out of its proper channels and we bleed, either internally or externally, life is endangered.

When the pink/red natural to the genitals...the passion natural to the genitals...stays confined within its proper sphere, the physical, health and new life result.

But when the pink/red of the genitals rises above its natural level, firing up the emotions, this intensifies the purifying work that must be accomplished, for the body to survive, by the heart and the lungs.

When the pink/red natural to the genitals...the passion natural to the genitals...rises even higher, past the emotional level to dominate the mental level, the head, and expresses itself on the face, it becomes a matter of anger and shame.

When red bursts forth onto levels where it doesn't belong, life is endangered. All signals scream STOP! Keep red...anger/sexual feeling...in its proper place.

* * * * *

With both form and color, our bodies seem to be telling us to give up sex, or at least to be wary of it, to be careful to keep it in its proper place.

Why so much negative accent on an activity that is, after all, absolutely essential for the continuation of the human race?

Not all forward movement, all progress, is completely benign, completely risk-free, as those of us living in the age of nuclear power know only too well.

When man first tamed fire, this was an enormous leap forward. He could keep himself warm and comfortable...if he didn't burn himself up.

The invention of the wheel was another gigantic forward step...as long as the creature pulling the wheeled vehicle watched what he was about and didn't inadvertently initiate such quick motion he was run over by what he was pulling.

Ships took man roving on the high seas...and oftentimes capsized and sank.

Airplanes went soaring into the skies...except for those that reversed direction and went plummeting into a diving crash.

The worldwide communications explosion of the 20th Century has brought all of mankind much closer...so we can frustrate, anger, misunderstand and wage war against not only our near neighbors but our distant neighbors as well.

With nuclear fission we might be able to develop an energy-rich, drudgery-free future for ourselves...if we don't sicken ourselves with radiation or bury ourselves in nuclear waste.

Progress brings risk.

If man is the result of an evolutionary process, then one of the gigantic forward steps he made was in the development of his feet. His thumb-like big toes moved in to where they were parallel with the other toes. In this position the big toes served to push-off and lengthen the stride. The heels came to rest on the ground to help bear and balance his weight. He developed "shock absorbers," a long, springy arch in each foot.

These changes are believed to be, by those who believe in the evolutionary process, one of the most significant changes in our anatomical evolution.

If we did not evolve but were created substantially as we exist today, then our designer gifted us with unique feet. There are 206 bones in our bodies. One fourth of these are in our feet.

The human foot is "a masterpiece of engineering and a work of art," according to Leonardo da Vinci. It is unlike the feet of other animals, distinctly our own, one of our most distinguished anatomical parts.

Our unique feet enabled us to become the only erect-postured biped on earth. Our hands, no longer needed to support body weight or to help with locomotion, were freed for countless sophisticated tasks. This made possible the accelerated development of our brains. It also changed coital postures, enabling man to indulge in positions unique to him.

Man's upright posture brought with it a change in what provoked sexual stimulation. The role of odor became less significant than the visual. Whereas olfactory stimuli were intermittent, visual stimuli were, or could be, continuous

Once man became upright, with his genitals fully in view, once sight rather than smell became the prime source of sexual stimulation, sex ceased to be periodic. It became instead...for the human animal only, a distinctly human experience...a year round, year in and year out, activity.

> *Fine and good, and all in the name of progress. But progress almost always brings risk. Are there risks inherent here too, in year round, year in and year out sex?*

Our bodies seem to tell us that yes, there are. *Why?*

COMMENTARY

In standing erect, man was pulling away from the physical, stretching up toward the sky. He was putting distance between his head, the mental, and the physical earth beneath him.

Sex is a physical activity. If man, exposed to constant sexual stimulation by his new posture, allows this stimulation to rule him...if he, in effect having drawn himself up from the physical, now drags the physical up after him into his head, allowing his emotional and mental life to be ruled by his sex drive...he is reversing the very process his upright posture set in motion.

If sexual obsession exists, it overwhelms and buries the very spirit that caused man to straighten up to reach for the sky in the first place. It defeats the very spirit that makes him human, man the unique, the upright biped, Homo sapiens, the one who is wise because he thinks.

* * * * *

So how to deal with this risk, how to avoid mucking up the whole damn thing, the great and glorious upright surge?

Perhaps through an examination of skin color we can gain hints of an answer.

4

Skin Color

◆

There's More Than One Way to Become an Upright Biped

> ○
> You profess to believe that "of one blood God made all nations of men to dwell on the face of all the earth"—and hath commanded all men, everywhere, to love one another—yet you notoriously hate (and glory in your hatred!) all men whose skins are not colored like your own.
>
> —Frederick Douglass

All normal skin color comes from four colors. The skin itself is white. A pigment in the skin adds a touch of yellow. A third color, black, is added through the presence of granules of melanin. Red is added by the blood circulating through the minute vessels of the skin.

White…yellow…black…red…mixed together give our bodies skin tones ranging from the very fair, almost pigment-less, to ebony black.

According to a recent news report, two scientists, Nina Jablonski and George Chaplin of the California Academy of Science, using data from a NASA satellite, have discovered why people mix up these basic ingredients into different colors. Variations in human skin color result from adaptation to the amount of ultraviolet light falling on different regions of the earth.

Differences in the mixture brought about an informal classification of man into five races based upon skin coloration, a classification no longer considered scientific but still lurking in some of our minds: *White, yellow, red, brown* and *black*.

Few things in man's history have been given the importance that skin color has. At the same time, the possible meaning of our skin's coloring has been given no consideration at all. We have seen without seeing. Skin color has been right under our noses for centuries, with rarely or never a thought to the richness of meaning it might have.

Black. White. Let's deal with these first.

The sun, a red ball of fire, slips below the horizon daily, plunging our world into darkness.

Man is a diurnal animal. Historically he has labored by day and slept at night. During the dark he slips into that little death we know as sleep. The blackness of night becomes, for much of mankind…though this is not universally true…the color of death.

Historically the night has not only been the time for rest, but also for such recreation as is available. This includes the activity that has long been considered man's favorite indoor sport, sexual intercourse. Black becomes not only the color for death but also one of the colors of sex.

This black association linking death and sex was expressed for centuries in the English language, when the words "to die" also meant to reach climax in sex.

While our life-giving yellow sun dominates our days, moonlight refreshes our nights. The silver-white moon, shining solely with reflected sunlight, appears to be a source of light but isn't.

In our world, if created by a designer who color-coded it, white becomes the color of illusion, of fraud and hypocrisy.

But brilliant sunlight, pouring down upon us from a yellow sun, is also white, the pure white light of the source of all life. This means that in our world, white is the color of purity as well as of illusion and fraud.

If our bodies tell us...if the coloring of the world tells us...if some deep underlying instinct tells us, the same instinct that told us to crawl up from all fours, to walk upright and reach for the sky...that sooner or later we must give up sex, or must at least outgrow the all too human tendency to be dominated by the lusts of the penis, how can we attack this problem?

Two different approaches suggest themselves, the two most common approaches used in attacking any addiction: Complete abstention or aversion therapy leading to a cure.

We can outgrow...outwit...the shrill insistence of the penis by turning a deaf ear, repressing our awareness of its demands, clamping a tight hold on ourselves much as a tourniquet cuts off blood supply. We can, with inflexible determination, decide simply to be so self-righteously pure that we will starve the damn thing, with its dangerous, never-ending lusts, until it dies of neglect. If this is the course we decide to take, what is more fitting than that we should head into cold climes to cool ourselves off and wrap ourselves in a lovely white skin?

We will take on the white mantle of purity and...still under the sway of that which we are determined to eradicate...become the purest hypocrites under the sun.

But we know who the enemy is...the blackness of sex...and will of course angrily, enviously, attribute to a black skin a sexual prowess fearfully beyond us, and we will rarely or never question our superiority to such blackness and our right to rule over it. We know that in opting for whiteness, for purity, we have taken the right course to rid ourselves of the dangerous black death.

Because the white-skinned search for purity through repression and denial infrequently meets with success, the hair of white-skinned people is curly-wavy more often than not, pubic hair undergoing a frantic straightening process that rarely really takes. Noses are narrow, the ins and outs pinched together...*breathe in repression, breathe out denial*... and mouths share the thin lips of the big apes, our near relatives who have sex only in season, strictly for procreative purposes, as we plan to do.

This is one way to deal with the dangers presented by the lusts of the flesh. Another, equally valid approach is through aversion therapy.

When faced with what seems an uncontrollable appetite, we may scorn the way of repression, knowing full well it leads to the stink of hypocrisy and a wholly unsupported belief in one's own superiority. We can decide instead to tackle it in exactly the opposite fashion, by feeding the appetite, hoping to reach, in fact to go well beyond, satiation, in this fashion bringing it to heel, an approach used today in many clinics as a hoped-for cure for alcohol addiction.

When white men first arrived in Africa, they were struck by the frequency of sexual expression among the native Africans, a frequency that persists to this day. Indiscriminate sexual activity is common in many areas of Africa as of this writing. Men and women alike indulge in sex with numerous partners, according to those who have observed the African sexual scene.

If aversion therapy...over-indulgence rather than repression...is the course we have decided to take, we would naturally not run from the heat but would embrace it, and we would stay put in the tropics. In recognition of our willingness to battle it out with our lusts, we would openly, courageously, wrap ourselves in the cloak of what we are challenging, that is, we would wrap ourselves in a lovely black skin.

Because this is a decision made on the mental level, however unconscious we are of the process, naturally we would cover our heads with a kinky, woolly black hair, pubic-like hair, the sexiest of all sexy hair.

Broad, flaring nostrils will show the world...*for those who have eyes to see*...that there is a marked diversion in the ins and outs of our quest, that we are inhaling life in huge gulps of passion in order to expel it again.

Our mouths, though on a mental level, connect directly with the physical, so to acknowledge the course we are following, we will give ourselves wide, deliciously full lips, evolving noticeably beyond the thin lip formation of the big apes and our hypocritical white-skinned brothers.

A second approach to the same basic problem, how to avoid dragging up to the mental level...permanently...the physical world we are standing upright to free ourselves from.

In general, though in some instances this correlation breaks down, those with white skins have lived in the colder climes and those with black skins in the warmer.

Denial/repression is a colder phenomenon than indulgence. We are never likely to see a bumper sticker that reads, "Cold to trot." When we get worked up, we get heated up. Sex is warm, heading toward hot. Warm air rises. The penis rises. Denial pushes downward toward the cold.

Denial=cold=white. Indulgence=hot=black.

Those who decided to try to lick the sexual challenge through denial/repression left sub-Saharan Africa and headed north, toward ice and snow. Those who decided to try to lick the problem through indulgence leading to aversion stayed put, in Africa, under the heat of the equatorial sun.

Or possibly those who headed north were persuaded, through the coldness of their world, that denial was the way to go, while those who stayed put were persuaded, through the heat of their world, that indulgence/satiation/aversion would work better.

It seems safe to say that, for the majority of mankind, neither of the above approaches has yet worked. As of this writing, both races stay mired...on the mental level...in the physical/sexual world, with the white, as is the way of the self-righteous hypocrite, still feeling envious and in consequence superior to the black, expressing this by demeaning the black to prove to him the error of his ways.

This white-black "brotherly" relationship parallels quite closely, in possibly an informative way, two approaches to the breaking of alcohol addiction. One alcoholic, proud possibly to the point of arrogance, may attempt to quit cold turkey, without help, through denial, an attempt that may or may not work. Another, having seen or experienced the failure of the denial approach, may seek help in a clinic or hospital, where he may be treated in some such fashion as this:

He will be given all the alcohol he could wish for, one drink after another after another, until he is too sick to want any more. Then he is forced to drink even more while placed in front of a mirror so that he is treated to the sight of himself vomiting. For this "help" he will pay quite handsomely…as the black race has always been forced to pay dearly for the "help" given him by the white.

Neither black nor white appears in the rainbow, and in the strictest sense neither is a color. White reflects…rejects…all light rays. Black absorbs…accepts…all rays. As suggested above, white is a denial, black an indulgence. But what of those skins that have an actual color, skins that are neither black nor white?

Anthropologists today divide mankind not into the five races based on skin color current early in the 20th Century, but into three main races based on various considerations. These three races are:

Caucasoid—wavy hair, narrow noses, white to dark skin.

Negroid—woolly hair, broad noses, brown to black skin.

Mongoloid—straight hair, flat faces, yellow, brown or red skin.

The white and black we have already considered. Let's look at the yellow, brown and red.

Our sun is yellow-white, yellow-tinged, which makes yellow the color of seeing and awareness. But any human being who takes the time to be aware of his situation, who stops to think of himself as a microscopic bit of life on a small insignificant planet spinning around on its axis, at the same time whirling in an elliptical path around a distant sun which is only one star in a galaxy of billions of stars, with the entire galaxy moving through space at a fantastic speed, and this only one galaxy in a universe of billions; that he comes into life howling with his first breath and from there proceeds inexorably toward a certain fate, the ultimate cessation of his breath, beyond which cessation he may or may not still exist...all of this is surely enough to give anyone pause. As Hamlet mused, "Thus conscience does make cowards of us all!"

Yellow equals awareness. Awareness makes us cowards. Yellow makes us cowards.

Yellow assures a cautious, middle ground approach, an approach that has been lauded, quite rightly no doubt, as wise, the inscrutable wisdom of the east.

If we're not quite sure we trust a given situation, if we feel uncertain and suspicious...above all, if we are of a cautious, yellow-tinted nature...we look at our situation askance, that is, sideways, obliquely, not quite straight on. To express this cautious approach, we would naturally go to the eye. The epicanthus is a fold of skin extending from the eyelid over the inner canthus or corner of the eye, a fold of skin so common in members of the Mongolian race that it is called a Mongolian fold.

Those of yellow skin look at life askance through what is commonly referred to, by other races, as a slanted eye.

Brown is yellow darkened, yellow enriched with black. The brown skin takes a less cautious approach, but remains allied with the middle ground.

A red-tinted skin also expresses a less cautious approach, verging toward an open confrontation, the warrior who will stand firm and take on all enemies, on whatever terms are set, even if that enemy proves to be himself.

The Mongoloid race proclaims its awareness of the primary problem…the need to overcome the lusts of the penis…through black hair, but black hair that is fine and straight, sexless hair, a pronouncement that while there is an awareness of the challenge, the challenge has been met and overcome. Or will be met and overcome, given sufficient time. A cautious, "wise" approach.

There is no indication that the middle way has yet been successful any more than success has crowned the efforts of the white or the black. All races at this writing remain in the race, breathing, struggling, dying.

But though no race as a whole has yet succeeded in its quest, every generation brings innumerable individual breakthroughs in every race, beings in human bodies who have become truly upright, in every sense, men and women who keep the physical where it belongs, beneath their feet, their hearts, once awakened in love, beating strongly, steadily in their chests without the stimuli of constant lust-driven excitations, and their heads in the sky, their eyes calmly, easily on the stars. These men and women, regardless of the material circumstances of their lives, step forth into a bright, beautiful, rich new world that some instinct in each of us urges us to seek.

ODDITIES:

For over a thousand years, one half of all Chinese women had their feet bound, beginning in early childhood, in order to break the bones and turn the foot back on itself, a procedure that rendered these women all but completely crippled. This was done in order to win the devotion of Chinese men, who were literally enchanted by the tiny, misshapen, three to four inch "lotus foot."

The "lotus foot," according to some authorities, caused an outpouring of erotic feeling on the part of Chinese men of such intensity that the thousand-year love affair it inspired has never been equaled anywhere else on earth. Is there any possible explanation for this seemingly wild aberration?

COMMENTARY

If, deep inside, all of us have the feeling that we have to... *must*...break our ties to the physical, to the earth, and know in our hearts and minds that the penis symbolizes, more than any other organ, this unhealthy tie that must be broken, some degree of repression is apt to occur in our erotic responses. It could scarcely be otherwise.

But what if–*what if*...sexual response was suddenly no longer tied in so closely with the physical? Or if we could somehow persuade ourselves that it isn't? How our feelings would soar, how erotic we could become! A woman with tiny, tiny feet...no matter that this is an artificial condition, deliberately visited upon her; we'll forget about that...has little contact with the earth. Her tie to the physical has been reduced to almost nothing. She is all but angelic, ready to float away. Consequently, feelings lavished upon her, erotic though they may be...and observers say they were the most intensely erotic feelings ever recorded...are not all that dangerous. This is *not* a devotion to the physical through sex but adoration of an unearthly being in human form, or so the Chinese male was apparently able to persuade himself.

ODDITY

The word "banana" denotes a fruit. In slang, it is used to signify a comedian, particularly in a burlesque show. In the plural...to go bananas...means to spout nonsense, to go crazy.

COMMENTARY

The banana is yellow-skinned but penile in shape. On the face of it, that is a comic or crazy combination. Who on earth has ever been able to be intelligently aware of anything with his penis?

ANOTHER ODDITY:

The albino, that rare individual lacking normal color pigmentation, has, as a result of the missing coloration, extremely sensitive, weak eyes, eyes so oversensitive to light that the eyelids are kept partially closed in protection.

COMMENTARY

Lack of color closes the eyes. In those of us with normal color pigmentation, shouldn't the fact of color...in ourselves, in the great world around us...cause us to keep our eyes wide open?

5

The Laws of Heredity

❖

When It Comes to Our Genes, Are We Like Peas in Their Pods?

o o o o o o o o o o o o o o o o o o o o o o o o o o o o o o
Man is greater than a world...then systems of worlds; there is more mystery in the union of soul with the body, than in the creation of a universe.

———*Henry Giles*

According to news reports, the human genome has now been mapped. In June of 2000 two teams of scientists announced that each had produced a draft version of the human genetic code.

President Bill Clinton compared this achievement to Lewis and Clark's mapping of the American continent. He then added that this mapping was even more significant.

A revolution in health care is promised with this mapping. There will be improved diagnosis of disease, earlier detection of genetic predisposition to disease, and possible gene therapy in which defective genes will be replaced by healthy ones. There is also the hope of new and more effective drugs based on all the new genetic information.

This brave new world will not dawn for several decades, it is said, decades during which scientists will have a great deal of hard work to do. But in the end the change in health care will be revolutionary.

It is through our genes...our human genome or genetic code...that we receive inherited traits from our parents. If we wonder why we have Dad's big nose but not his broad, charming smile, why we have Mom's stubby fingers but not her long, shapely legs, why we have Grandma's weak chin but not her piercing blue eyes, Grandpa's sloping shoulders but not his beautifully shaped head, we should look to our genes, according to the science of genetics.

Until recently most experts in the field believed it took a hundred thousand genes or more to create and manage a human body, but this is no longer thought to be so. In early 2001 scientists announced that we humans need fewer genes than was earlier believed. Now the guess is that we have only thirty to forty thousand genes, which is only a few more than a roundworm has or a mustard weed.

Searching for a gene in the vast expanse of the DNA is the proverbial looking for a needle in a haystack, according to Dr. Francis S. Collins, leader of one of the two research teams who are mapping the genetic code.

Genes are transmitted to us in chromosomes, which are threadlike bodies that are found in a cell nucleus and carry the genes in a linear order. Every human being inherits a set of genes, contained in twenty-three chromosomes, from each parent. Cells in the human body are diploid, that is, every cell contains a double set of similar chromosomes.

When the body reaches sexual maturity, gametes...sperm or eggs...are formed. During the gamete maturation process, reduction occurs and the diploid chromosome number becomes haploid, that is, the gamete contains only one set of genes-bearing chromosomes. When the egg and sperm unite, each tosses in a single set of twenty-three chromosomes to the new organism, bringing the total back up to the standard diploid number of forty-six.

So there we are, inheriting from both Daddy and Mama. But how is it decided that we'll be cursed with Daddy's thick, heavy eyebrows instead of Mama's arched, perfect ones, or her pale brown, watery eyes instead of Daddy's clear, lovely blue ones? Why do we look as we do? Why do we have the talents we have? How did we happen to become ourselves, with the particular combination of facial features, body configuration and personality traits that we have?

Pioneer work in the field of genetics was done by an Austrian monk, Gregor Johann Mendel (1822-1884).

Mendel worked with pea plants that bred true to type. Pea plants that bore yellow seeds always produced other yellow-seeded plants, and green-seeded produced green-seeded. He wondered what kind of offspring would result from artificial cross-fertilization of true breeding plants.

If a tall variety of true breeding plant was fertilized with pollen from a true breeding short plant, what height offspring would result? Would the offspring be intermediate in height between the two parent strains? Would all be tall? All short? Or would some be tall and others short?

The actual result of his cross breeding was that all the offspring plants were tall.

Mendel used the word "dominant" to refer to the characteristic that appeared in the offspring, and the word "recessive" for the characteristic that disappeared.

He studied a variety of pea plant traits, cross-fertilizing plants that differed in these traits. The results of these experiments led to the formulation of the "Law of Dominance." When plants with contrasting traits are crossed, all of the offspring inherit the trait from the same parent strain. The other form of the trait, from the second parental strain, disappears.

It is not yet known how this dominance is achieved, that is, how one tall gene in a hybrid is as potent in achieving tallness in the pea plant as the two tall genes of a true breeding plant. To further mystify the matter, the fact that a given trait is dominant in one species does not mean it will be dominant in another. The color black is dominant over white in rabbits, but white is dominant over black in the Leghorn chicken. In some cases neither characteristic of a pair of traits can dominate the other and a blending of traits results. This is known as "incomplete dominance" or "blending inheritance."

Examples of complete and incomplete dominance in Homo sapiens are: Dark hair and eye color are dominant over light hair and eye color, but with skin color there is incomplete dominance, so that the offspring of black/white parents will show a blending inheritance in skin coloration.

After observing the dominance of certain traits in pea plants, Mendel was curious as to what had happened to the recessive traits that had disappeared. He had the offspring (hybrid) generation self-fertilize...and the recessive trait reappeared in the new generation, in unaltered form. The rates of the dominant to the recessive, in experiment after experiment, approximated three to one.

This led Mendel to formulate the "Law of Segregation." The recessive trait that is hidden in the hybrid generation is not lost since it reappears in subsequent generations.

Further experimentation led Mendel to formulate the "Law of Independent Assortment," a law that states that every characteristic is inherited independently of every other characteristic.

The above Mendelian Laws of Heredity are today considered principles rather than laws as it has since been found that there are as many exceptions to each rule as there are cases that follow the rule.

Others besides Mendel began to develop "laws" of heredity, and in 1900 *genetics*, the science of heredity, was born.

A great deal of work was done with the tiny, red-eyed fruit fly, Drosophila, and a great deal learned about how heredity works through these studies.

To determine how heredity works in man poses a somewhat more difficult challenge than to figure out how it works in pea plants or fruit flies. An experimenter can cross breed pea plants at will, as Mendel did.

Fruit flies obligingly reproduce at a rate of twenty-five generations per year and hundreds live comfortably in a quart bottle.

It is not possible at present to breed people at will, to suit the experimenter, and the twenty-plus year gap between generations makes it difficult to follow the appearance/disappearance of given traits through several generations even if breeding were under laboratory control. Extrapolation from pea plants, fruit flies and other sources, plus thousands of years of observation of the human condition..."Johnny is certainly the spitting image of his Dad!"...have led to the belief that man is, in general, subject to the same principles or laws of heredity that have been proven to be operative throughout much of nature.

How valid is this extrapolation from pea plant and fruit fly to man?

The relatively young science of genetics has formed close ties to several other disciplines. From Mendel's day, it has leaned heavily on mathematics. Statistics, probability and mathematical analysis are called upon. Genetics has also formed a close alliance with *cytology,* the study of cell structure; has made use of x-rays, ultraviolet light, atomic energy, and other forms of radiation; and has also formed a partnership with chemistry. Biochemical genetics seeks to understand the effects of heredity on the chemical reactions that occur in all living bodies. The intent is to find out how genes control the fundamental chemistry of life.

But before we attempt to learn *how* genes control our chemistry, shouldn't we at least ponder and try to test the question as to *whether* they do? Possibly a valid and rewarding study would result if we were to flip over the scientific table and test out theories instead on how exactly the chemistry of life...the chemical reactions occurring in our bodies...exerts control over our genes.

According to current thinking, the genetic makeup inherited by any given individual is largely a matter of blind chance. If a brown-eyed parent happens to carry a recessive gene for blue eyes, it is a toss of the die as to which gene...the dominate one for brown eyes or the recessive one for blue...will turn up in his offsprings' genetic pot. The same goes for a vast assortment of other traits that are said to be inherited. Blind chance is further modified by accident, and again by more chance. All in all, the entire hereditary package is extremely chancy.

Mendel's original "law" that each characteristic is inherited independently of every other has failed to stand up upon further experimentation. It has been found, instead, that traits are often inherited together, due to a phenomenon known as *linkage*.

While organisms may have hundreds of thousands of inheritable traits, the number of chromosomes bearing the genes is limited. Therefore each chromosome must carry many genes. The genes on one chromosome, linked together, traveling together, are inherited as a package...unless, by chance, *crossing over* occurs.

Crossing over is a breakdown in linkage and no one knows why it occurs. Current thought attributes it to chance.

The name expresses what happens. The paired chromosomes of a diploid cell...a cell containing two sets of genes...duplicate themselves, and the two newly formed chromosomal strands come together and twine about each other. During this entwining, crossing over may, by chance, occur. Before the chromosomal strands pull apart again, genes from strand A may cross over and join strand B, while B genes join strand A.

The effect of crossing over is that even genes that are "linked" together in a single chromosome may not remain together in the new generation.

There are also gene changes, ie, mutations.

Genes are believed to be extremely stable, but mutation...sudden change...can occur. Genes are subject to change through chemical means, also through the use of x-rays or heat. H. J. Muller, Nobel Prize winning geneticist for his work on mutations, found that many of the mutations that he produced artificially were similar to those occurring naturally.

There is a wide variation in the mutation rate for different genes. Researchers have concluded that males are responsible for more mutations than females. This holds true for mutations that are unfavorable, introducing new diseases into the human family, and for favorable ones that allow evolution to move forward.

In addition to mutations, various other changes and disasters can occur to chromosomes on their way to supplying you with your genetic inheritance. Pieces of chromosome can break off during gametic formation and get lost. The lost pieces can attach themselves to other chromosomes. Heredity is altered as some gametes will be deficient...missing the chromosomal piece...while others will have duplicates...they got stuck with what was lost.

Sometimes two chromosomes that are not members of the same pair can exchange pieces (this differs from crossing over, which involves two chromosomes of the same pair). Sometimes chromosomes fail to separate during reduction division. Other chromosomes may wind up upside down.

As we become acquainted with all the disasters that can befall our genes...all the possible ills the poor things are heir to...it begins to seem somewhat miraculous that any of us comes out in a more or less understandable shape, resembling, to some extent at least, our parents and our siblings.

Why do I look as I do? Why do I have the talents I have? How did I happen to become me, with the particular combination of facial features, bodily configuration and personality traits that I have?

Inherited traits are determined by genes, the experts tell us, and chance determines which genes we will get. Chance also determines how battered and bruised...properly placed, misplaced or lost, upside down or right side up...these genes will be.

Let's apply this to imaginary case histories to see how it works.

Height is considered to be an hereditary trait, influenced/modified by environmental factors, so:

Six-foot, two-inch Harry marries five-foot, two-inch Mary and they produce three sons.

The oldest son attains an adult height of six feet, five inches, three inches taller than his father. Genetics tells us how this happened. It happened by chance.

The middle son grows up to be six feet, two inches, the same height as his father. Genetics tells us how this happened. It happened by chance.

Their youngest son attains a height of only five feet, ten inches, the same height as his mother's tallest brother. We know how this happened. It happened by chance.

Eye color, hair color, and facial features are also considered heritable traits, though they too can be influenced/modified by environmental factors.

Handsome Hans marries a plain Jane and they produce three children, two boys and a girl.

Their first born, a boy, "takes after" his father and is extremely good looking. Their younger son takes after his mother, not his father, and is considered to have greater inner beauty than outer. The youngest, a girl, has such a mix of features that she doesn't look very much like either parent. She is thought rather unattractive as a child but in adolescence unexpectedly blossoms into a genuine beauty. All of this happened through heredity, that is by chance, according to the science of genetics.

Whether, with our green eyes and auburn hair, we most resemble Dad's brother Tom or Mom's cousin Beatrice, if Dad supplied a human sperm and Mom a human egg, the chances are close to one hundred percent that if the fertilized egg survives, it will mature into a recognizably human body, with the standard operating equipment. Fertilized human eggs grow into human babies, not into baby chimps or orangutans. This is assured by genetic inheritance and, in distinction to superficial traits like stature, coloration or facial features, there is very little that is chancy about it.

Relative to this, recent research has discovered that the genes responsible for the laying out of an embryo's anatomy are very densely clustered and contain no "junk" at all, so "junk" has little chance to interfere with their operation.

This "junk" that isn't present within the densely packed gene cluster responsible for the complicated, all important task of laying out the anatomy but is abundantly present elsewhere, constituting up to 98% of the DNA, will be discussed in the next chapter.

There has been a great deal written in recent years about DNA (deoxyribonucleic acid), RNA (ribonucleic acid), and the cracking of the genetic code that has resulted in gene splicing and the creation of new life forms in the laboratory. DNA, usually found in the nucleus of the cell, is seen as the master builder, in charge of the blueprints. RNA, produced by DNA in the nucleus, forges out into the cytoplasm (the protoplasm of a cell exclusive of the nucleus) carrying the blueprints, "instructions" from the boss DNA as to what kinds of proteins are to be built. Building proceeds apace and growth results, genetically determined.

If the DNA blueprint calls for an E *coli* bacteria cell to be built, that's what is built...unless laboratory experimenters intervene and alter the genetic code. This gene splicing/creation of man-made forms is currently being done on the bacterial level, with enormous implications for all forms of life.

Is man as vulnerable as the E *coli* bacteria to forces external to himself manipulating his genes?

Is he as vulnerable to the operation of chance as science claims he is in his genetic inheritance?

To what extent, if at all, is man at the mercy of his genes?

Mendel could work out "laws" of heredity by experimenting with pea plants, but can these same "laws" be applied to man?

The tiny, red-eyed fruit fly, Drosophila, can be bred generation after generation in the laboratory and a great deal learned about its hereditary patterns, but do these findings have any applicability when it comes to human beings?

The genetic scientists have answered that yes, heredity in man does follow much the same pattern as that observed in pea plants, Drosophila, and in numerous other animals and plants. Everything found in the inheritance of plants and animals is also found in human inheritance, we are told. There is evidence in human heredity of dominance, blending, segregation, recombination, independent assortment of traits, linkage of traits, sex linkage, mutations, and so forth.

In the absence of direct laboratory experiments, this evidence has largely been gathered through a statistical attack, information drawn from family pedigrees, medical records, prison records, from the studies of twins and random populations. All of these studies have led to the conclusion that human beings, like pea plants or red-eyed fruit flies, are the product of their genes and the interaction of their genes with the environment.

Is this a valid conclusion?

Possibly, but let's not forget that lots of things work differently for pea plants and fruit flies from the way they work for man. Consider this series of statements:

Pea plants have not been endowed through genetic inheritance with the wings needed for aerial flight. Therefore pea plants will never fly.

Man has not been endowed through genetic inheritance with the wings needed for aerial flight. Therefore man will never fly.

Fruit flies have not been endowed through genetic inheritance with the large, powerful, ocean-adapted body of the whale. Therefore, fruit flies will never roam the high seas as whales do.

Man has not been endowed through genetic inheritance with the large, powerful, ocean-adapted body of the whale. Therefore man will never roam the high seas.

Pigs have not been endowed through genetic inheritance with the body structure built for great speed. Therefore pigs will never move across a land surface at a speed faster than a horse or a deer.

Man has not been endowed through genetic inheritance with a body structure built for great speed. Therefore man will never move across a land surface at a speed faster than a horse or a deer.

Clearly man has accomplished things undreamed of, as far as we know, by the pea plant, fruit fly or pig. Either through genetic inheritance or through some other means, he has been endowed with a trait not shared with other life on this planet, or not shared in equal measure, and that trait is creative intelligence.

Creative intelligence has freed man from many of the limitations imposed by his genetic inheritance. Due to this creative intelligence, man is not completely at the mercy of his genes.

Let's risk taking this one step farther and suggest this hypothesis:

HYPOTHESIS:

Man is in no way at the mercy of his genes.

Dominance–blending–crossing over–mutation–deletion/translocation–segmental interchange–inversion–non-disjunction. Let's forget all this inferred genetic behavior, put a creative finger on the trigger, and suggest the above hypothesis not in negative phraseology but in clear and positive terms.

HYPOTHESIS:

Man has free will, and this free will extends even into his genes. Unlike the pea plant and the red-eyed fruit fly (or possibly not unlike them; who knows?), man creates his own body to suit his own needs.

* * * * *

The premise that we have this amount of freedom...this measure of mind-blowing creative power...will be explored in the next two chapters.

6

Heredity Versus Environment

❖

Nature or Nurture: Which Force Forms Us?

o o o o o o o o o o o o o o o o o o o o o o o o o o o o o
And God said, Let us make man in our image, after our likeness.

—*Genesis 1:26*
Bible, King James Version

During the early years of the 20th Century, there was an ongoing controversy as to which force, nature or nurture, played the bigger role in determining who and what we are. It was heredity *versus* environment.

Today genetics is defined as the "science of heredity," dealing with the resemblance and differences of related organisms resulting from the interaction of their genes and the environment.

Heredity, science tells us, is determined by an interaction between the former combatants, genes, the transmitters of heritable traits, and the environment, the aggregate of surrounding things, conditions or influences. In effect, genetics has now become an all encompassing package, heredity interacting with environment, heredity plus environment.

On the face of it, the word *genetics* would seem to pertain to genes, not to such environmental conditions as level of temperature, exposure or non-exposure to light, proper nutrition, etc. In practical application, however, it was found to be difficult if not impossible to unravel the role of the genes from the influence of the environment. The two worked too closely together.

Improper nutrition could cause stunted growth regardless of the genetic inheritance, and a fault in a gene or a poor environment could result in similar or identical faults in development. Identical malformations can result from a defective egg in a good environment or from a healthy egg in a bad environment.

If genes are this dependent upon the environment, if an unfavorable environment can cause malformation and good genes are no guarantee against faulty development, do we inherit anything at all worth considering through our genes?

Yes, say the genetic scientists. Although, due to the crucial importance of environmental factors, we can not be said to inherit characteristics or traits, we do inherit *capacities.*

Capacity is an actual or potential ability to do something. Whether the capacity is actualized or remains a potential depends upon the environment.

On the dictionary page, genetics, "the science of heredity," has taken over the enemy territory, the environment. But when we look at how this developed we find that the exact opposite actually occurred. Heredity declared itself subject to environment, the genes being stripped of power to where they are no longer believed to pass on characteristics or traits, but simply an "ability to react."

To react to what? To the environment.

Through our genes we inherit potential. How restricted or unrestricted is this potential? In interacting with our environment, how much choice, if any, do we have as to what potential will be actualized?

Those who believe in conventional genetic theory offer us a narrowly restricted range of potential. For example, geneticists tell us that hair, eye and skin coloring are all inherited. In the human eye, brown is said to be dominant over blue. This means, according to conventional theory, that if a gene for brown eyes is passed along, tossed into the genetic pool, the offspring will have brown eyes...provided the environment is favorable enough to allow for normal development. The environment can thwart normal development or it can allow normal development to occur. Nothing more. Its role is strictly that of a possible spoiler. The power to determine development, in a "normal" environment, still resides in the gene.

Admit the influence of the environment and the new organism is still seen as having no power of choice. Instead, in conventional theory, it is seen as under the control of two factors rather than one, as subject to the mastery of its genes and also the vagaries of its environment. At no point...in conventional theory...is it believed that the organism can take a stand and decide for itself: "Hey, Kid, we've got genes here...blueprint plans here...for brown eyes or green or blue, so...what do you say...let's have blue eyes, okay?"

This kind of fairly unrestricted potential is ruled out by genetics, the science of heredity.

What is the immediate environment of the human gene?

DNA (deoxyribonucleic acid), the transmitter of hereditary characteristics, has been shown to be the master builder, the brains of the outfit. DNA is found in each of the cells of a human body. In fact, it exists there in an embarrassingly large amount.

The immediate environment of human DNA is the cell it finds itself in. The cell is a tiny part of a complex, magnificently organized structure with an intricate network of nerves, a heart to pump life-sustaining fluid through it, and a brain capable of storing a vast amount of knowledge. This brain gives rise to or is associated with...though no one knows exactly how...a mind capable of all kinds of wonderful things, thoughts, feelings, logical analysis, day dreams, night dreams, and the creative imagination that has allowed its owner to burst the restrictive bonds of its own genetic inheritance, to soar through the air, roam the high seas, speed across the land, in effect to master its environment in almost every way. Yet this powerhouse of a body, with its attendant magical mind, is seen as all but completely at the mercy of the DNA in the nucleus of its cells.

Have we no potential at all other than that offered by the DNA in our cells?

Research has shown that DNA does not work the way it was originally believed to work.

In the *Escherichia coli* bacteria, where DNA was first observed at work, biologists watched as the DNA duplicated itself segment by segment in the RNA, then the messenger RNA left the nucleus to carry the genetic message out to the cellular machine where the genetic command was executed. It was all very neat and satisfying, and biologists assumed that this was how it worked in all living things.

Subsequent study, however, brought about a revision of thought as it was found that DNA in the more highly organized forms of life did not work in this same tidy, non-troublesome way.

In more highly organized forms of life, DNA did not duplicate itself segment by segment in the messenger RNA. Instead DNA was found to contain numerous unused strands, strands that the biologists promptly labeled, to start with, "genetic gibberish." The messenger RNA matched up only with "sensible segments," ignoring the "nonsense interruptions," ie, the "genetic gibberish", before departing the nucleus. These so-called "split" genes...DNA split between *exons*, which are sensible, and *introns*, which are not...have now been found everywhere, in every species examined, with few genes not sharing this fate. Some genes are splintered in spectacular fashion, broken up into as many as fifty parts.

If we hope to find some freedom in human inheritance, we may now have found it in all this unused DNA, all this non-coding, non-actualized, *potential genetic material.*

So resistant to the possibility of choice in heritable traits are the genetic research scientists that this unused DNA...the non-coding, unused potential...is referred to as "junk." Only recently have they begun to see that possibly this "junk" has, or at one time had, some biological function, either in the evolution of the species or in its creative work today. But never is it suggested that possibly the unused segments indicate individual choice.

This blindness persists despite the fact that the exons are usually much shorter than the introns, the useful exons averaging around one hundred to three hundred genetic letters while the introns, the non-coding stretches of "junk," average around one thousand letters. Some introns have as many as ten thousand letters, all of as yet unknown use, according to the geneticists.

This seems a curious situation indeed. The human cell has around a thousand times more DNA than a bacterium. No one is sure what all this extra DNA is for. When it is found that only a small part of the DNA is expressed...that exons, the coding letters, are usually much shorter than intragenes, the non-coding segments...no biologists suggest that possibly this has something to do with the wide range of human potential, or with the working of the wide ranging human imagination.

> *If all this unused DNA is really waste material...gibberish, containing nothing but nonsense, helpful only in some minor and as yet not-understood way...why do we humans have so much of it and E. coli bacteria none at all? The human body is usually found at or near the top of the heap when it comes to the good, usable things in life, not loaded down with a lot of garbage. Why does each of our cells have as much as a thousand times more DNA than a bacterium has when most of it is "extra," ie, known as "junk"?*

Let's consider two parallel situations.

An experienced builder of mousetraps sets out to build one. He makes a list of the materials he needs, gathers this material, and builds his mousetrap. He winds up with a well-built mousetrap and no excess material whatsoever, all neat and tidy in imitation of the way DNA works in the E *coli* bacteria.

A second builder plans to build a cathedral but he is unfamiliar with the site upon which he will erect his structure. He doesn't know beforehand whether he wishes to build it with brick, stone, marble, wood, stucco or steel, and won't know until he is actually on the building site. However, once he is on the site it is too late to order additional material. Therefore in making out his list, ordering and gathering his material, he includes every possible building material along with the bonding agents needed to work with each, nails, mortar, or whatever, also all tools required for each and every type of construction. All of this adds up to an enormous amount of material, most of which he knows going in won't be used.

He arrives at the construction site, looks everything over, measures, considers, possibly meditates, and then decides, "I shall build myself a cathedral with steel walls, a great deal of glass, and stone decorations." He orders blueprints, then extracts from his huge pile of materials the ones he has decided to use and turns his back on everything else. He will build the cathedral that his mind...his free will...his creative intelligence...has instructed him to build.

The material he has left unused is for the moment extra, extraneous to his purpose, but it remains building material whether or not it is being used. Quite possibly, as he builds his cathedral, he will alter his plans somewhat and call upon material not called for in his original blueprints.

Due to his careful planning, the potential for changing his structure is always there, for he started out covering every base he could think of, ordering enough material for every possible contingency. As he could not possibly know ahead of time what kind of cathedral he would want to build, no one should fault him for being extravagant, for ordering more than he needed. He did *not* order more than he needed. He simply did not find use for all that he had. But maybe tomorrow, the week after next, or next year he will find use for it. Building the cathedral is his one task in life and he will never really be finished until he dies.

Hair coloring is believed to be hereditary, one of the many traits ruled by our genes, yet few babies of Caucasian ancestry keep the hair coloring of the soft down covering their skulls at birth. For many the blond hair of early childhood tends to darken with age.

As a very young child I was towheaded, with fine straight hair that was almost white. Over the years my hair became golden blond and thickened to become wavy-curly. During my teenage years it darkened to light brown, then to darker and darker brown. By the time I was forty it was so dark a brown as to be almost black and was streaked with gray. By the time I hit fifty it was completely gray. Since then the gray has lightened more and more toward white. It has not yet lightened completely back to the towheaded white of my babyhood but I expect it soon will. That's quite a range of hair coloring for anyone to "inherit."

Eye coloring is another of the traits believed to be hereditary, determined by genes. Most Caucasian babies are born with blue eyes. For many of these babies, the color will change soon after birth. I was blue-eyed into my mid-teens, then my blue eyes changed to green and that's the color they have been ever since.

The blue-eyed blonde I was at twelve matured into a nineteen-year old, green-eyed brunette. How is this accomplished under conventional genetic theory?

First of all, according to such theory, I inherited from each of my parents, one green-eyed, the other hazel-eyed, recessive genes for blue eyes. This happened by chance. I could as easily have inherited a gene (or genes) for green eyes from one, hazel eyes from the other, in which case I would have had hazel eyes, a light brown eye, brown being dominate over blue or green. But the recessive genes won out and I was blue-eyed. The gene for blue eyes carried a built-in instruction (apparently) to change my eye coloring to green sixteen or seventeen years down the road. The DNA that contained the genetic instruction to give me blue eyes to start with, green eyes later on, was a short, sensible, coding segment in a long strand containing mostly non-coding "junk."

The genes controlling hair coloring had to be given more elaborate instructions: Start white, change to bright blond, darken, darken, darken, then lighten, lighten, lighten back, in time, to the original white. These genetic instructions, too, came from exons, coding segments of the DNA strand, with "nonsense" stretches on either side.

Is this really how it happened? I'd like to consider another script.

SPECULATIVE GENETIC SCRIPT:

Some mind or spirit wanted to build a human body. Once the decision was made to cover the body with a white skin, it was decided to start with blue eyes. So along the strand of DNA containing instructions on how to color the eye...instructions for eyes of blue, blue-gray, blue-green, green, hazel or brown...it zeroed in on the blueprint of its choice, ie, an eye colored blue. This was the particular blueprint wanted so the other blueprints...for green, gray, hazel, or brown eyes...are not called forth but are left for later consideration.

The blue eyes satisfactorily express what the mind or spirit in the body wants to express for some sixteen years. But then, with a change in circumstances, green eyes become more appropriate. The word comes down that in recreating the molecules that form the eye, there is to be a switch over, as quickly as can be comfortably done, to a different blueprint on the DNA strand, to the one that calls for green eyes and express that instead. Green proves to be satisfactory, expressive of what needs to be expressed, and no further changes (to this point in time) are ordered.

With hair coloring, the same procedure is followed. As the new hair grows in, a shift is made along the row of blueprints that determine hair color so that the hair coloring remains expressive of what the mind or force wants expressed. To begin with one small segment of DNA is expressed, then in time a shift is made and a new blueprint called forth. One shift after another occurs as the body ages and heads toward the grave.

In both of the above cases, it is taken for granted that DNA is the master builder, in charge of the blueprints. The first, the "scientific" explanation, calls for only a small part of the DNA to be expressed while arguments rage as to why we are loaded down with so much unused "gibberish." The second theory, which allows us much greater freedom from the tyranny of our genes, assumes that *all* of the DNA is sensible and able to be expressed, that every strand contains its allotted portion of blueprint whether or not every blueprint is called forth and constructed.

The crucial point of divergence is that in the former case, the DNA is viewed as its own master, in charge of itself. In the later case, while the DNA is seen as in charge of the building process, it is viewed as being strictly at the command of a force or will beyond itself. The genes are reduced to the role of underlings, lacking self-determination.

COMMENTARY

The genetic scientists have possibly been so enthralled watching DNA, the master builder, at work, marveling as the Boss synthesizes messenger RNA to match the chosen blueprints and sends this RNA on its way out to the cell, that they may have forgotten that behind most builders stands another, the one who visualized the structure to begin with and drew up the blueprints, or had them drawn up, the one who may revise here and there to suit unexpected contingencies: *The architect.*

It is suggested that, in the building of the human body, free will, inspired by emotion, guided by the creative intelligence, is this neglected, so far undetected presence, the architect.

It is further suggested that it is the architect...not the DNA, with its mass of blueprints...who decides exactly which blueprints will be called forth and built, who has, in effect, the final say-so on all matters of construction.

HYPOTHESIS:

> *Man has free will, this free will extends even into his genes, and he builds his own body to suit his own needs. From first to last, from the womb to the grave, he is his own architect.*

* * * * *

Is there any proof that man's free will has this much power, that such an architect of the human body exists?

No, there is no proof at all. It has never been seen or otherwise detected and can only be inferred. But let's not forget that the human brain has been observed, tested, even surgically operated upon, yet the mind, which is viewed as in some way inhabiting the brain, has never been seen or in any other way detected either. Does this prove that the human mind does not exist?

Envision: To contemplate in imagination, especially as coming into being. The suffix–*ment*:...denoting an action or resulting state. There apparently is no such word as *envisonment*, but possibly there ought to be, with this definition:

> *The process or action by which the human will brings into material being its physical form, the human body.*

How this process or action of envisionment, as defined above, might work is considered in the next chapter.

7

Envisionment

◆

If Junior is "The Spitting Image" of Dad. Is He Sincere—or Sincerely Repressed?

o o o o o o o o o o o o o o o o o o o o o o o o o o o o o
Example has more followers than reason. We unconsciously imitate what pleases us, and approximate to the characters we most admire.

—*Christian Nestell Bovee*

We all know that the human animal is prone to imitating others. Little girls don the high-heeled shoes of their mothers, smear on makeup, and parade around as miniature "Mommies." Little boys slip their feet into Daddy's work boots, pick up his lunch pail, and stride manfully through the house. "Imitation is the sincerest flattery." Nathaniel Cotton.

But long before the young girl or boy is old enough to take on the clothes, manners and beliefs of parents or other loved ones, a human body has been forming, a body that began within the broad framework of a human mold but with little specification beyond that.

Most newborn babies look like other newborn babies. The parents of a newborn who have not yet seen their offspring, if allowed to walk through a hospital nursery wherein are kept fifty or even twenty newborns, would be hard put to identify their own simply on the basis of appearance. Immediately following birth, newborns are name-tagged to try to make sure the new mother takes home the child to whom she gave birth. Newborns are also foot-printed as another hedge against inadvertent mix-up. The newborn Jones baby looks far too much like the newborn Smith baby and too little like his parents for much reliance to be put on appearance.

Despite this lack of resemblance at birth, the Jones boy quite possibly will grow up to be a carbon copy of his father. Or perhaps he'll grow up to look remarkably like his mother instead, while the Smith boy will quite possibly grow to strongly resemble one or both of his parents. How does this resemblance come about?

Standard genetic theory assumes that heredity causes it, that our genes program us to look like our progenitors. If we fail to follow this pattern and grow up looking more like the people next door, genetics readily supplies "logical" reasons for this: Dominant versus recessive genes, plus or minus all the oddball hazards the genes are heir to. Any given organism can always be a throwback to unknown forebears...still through the genes...or maybe a couple of handy mutations occurred.

Free will may come into play...does come into play...when it comes to our posture, our stride, our facial expressions, our speech patterns, our religious beliefs, and our general behavior. It may determine whose clothes we dress up in when we're little and whose hopes and dreams we take on as we grow up, but it is assumed not to have any part in forming the underlying physical structure itself. In this area, in conventional genetic theory, the genes are still in control, in interaction with the environment.

But one of the major forces, if not *the* major force, in the immediate environment of human genes is the human mind with the human will.

HYPOTHESIS:

The mind/will is the architect that decides which genes to use as it builds its body.

The human infant...as she opens herself up to her external environment...soon realizes how utterly dependent she is upon those around her. As she learns to both fear and love those who take care of her, she interacts with her inner environment...her mind/her will...by taking on their facial features, their hair and eye coloring, their speech patterns, their expressions, their prejudices and their dreams.

First the physical body itself, then all the rest.

This is a lifelong process, ending only when life in the physical body ends.

As we find new loves and painfully uproot the old, as we become joyful or disillusioned, as we unearth our buried fears and deal with them or leave them festering deep within, we modify our bodies in keeping with our needs. A daughter who may in no way resemble her mother physically at age fourteen may grow to be almost a carbon copy of Mama by the time she is fifty. A round, broad, adolescent face may change, as mine did, into a long, narrow, middle-aged one. Or a narrow face may grow to become a broad one. An eighteen-year-old daughter who, in her mother's wedding dress, could have posed for her mother's wedding pictures, may, by age fifty, have traveled far from the original. We reshape our faces as we go, slowly molding them to express what's in our hearts, revise our facial features much as we broaden and change our vocabulary, alter our stance, our walk, our way. If we rebelled in childhood and at fifteen were a mutant..."How did she ever get into this family?"...by age fifty we are often very clearly back in the fold. If we lovingly followed our hearts in childhood so that on our wedding day we were Mama reborn, new loves may lead us far astray before we die. As our feelings emerge into the light and are expressed, as our fears are faced and die away, as our love contracts or expands, our wills reflect this and our bodies change. Life is not a static, gene-controlled process, but a vital flow of feeling and will.

* * * * *

Q. If we human beings have the power to create ourselves in whatever form we wish, how is the creation accomplished?

A. *The same way most things are accomplished: Through use of the raw materials at hand.*

Science has shown us that we inherit chromosome-bearing genes, that the DNA in the nucleus of our cells is the master builder, sending out orders as to which proteins are to be built. We know that DNA exists in what seems to be a greatly surplus amount. The greater part of this DNA is what the scientists call "non-coding," that is, not replicated, but this non-use may be temporary. Quite possibly we can call up "non-coding" genes as we please and turn them into "coding" genes.

It has been shown in the laboratory that genes can be altered through chemical means, also through electrical means and through the application of heat.

Strong emotions in human beings bring about proven chemical changes in their bodies. Our brains work on electrical impulses. Bodily heat can vary depending on physical and emotional health.

Even if genes "rule" over us, we possess ample means at hand...chemical, electrical, temperature level...to alter them, to put our heart's imprint upon them, to bend them to our wills.

COMMENTARY

Possibly this is how we create our own bodies:

The cerebrum, the largest, most recently developed, most human part of the brain, has two hemispheres. The left rules over abstract reasoning and the verbal, the right over feeling and the spatial. The left is logical, the right creative.

If, as an infant, we adore Daddy but don't yet have the words to express this, or if the feeling is too big for words ever to express, the right brain may be stimulated to express this love creatively, through imitation, the sincerest flattery. We take on the features and general bodily build of the one we adore.

On the other hand, if the feeling we naturally have for one of our parents is repressed, through fear, pushed out of the verbal left brain, pushed completely out of thought, it must go somewhere. Surely love is too powerful a force simply to die away. So the unthought feeling may splash over into the creative right brain, where it will be expressed in bodily imitation. Murder may not always out, but love will, if not in thought and words then in face and form.

We can "take after" one parent more than the other so that an excess of love can find expression, or as the sole expression of a love we are too frightened to face.

Throughout our lives, as we work through our fears and allow buried feelings to surface, we modify our coloring, our facial features and our forms to express these changes. Nothing is left to chance. Rather, the free will architect has the final say-so on everything, from the instant of conception until the plunk of dirt on top of the coffin as the physical body gets lowered into its earthly grave.

* * * * *

But what of adopted children who may grow up to resemble birth parents they have never seen rather than the adoptive parents with whom they have lived intimately all their lives?

A possible explanation: All adoptive parents and most adopted children know the truth of their relationship, that they are not blood kin.

If the only link you have with an unknown but secretly rather longed for past is a tarnished gold locket, one that you don't have the means to open but which may contain a greatly valued treasure, you would surely take extraordinary care of this locket hoping that someday someone would come along able to open it for you and reveal its contents.

Oftentimes the only link the adopted child has with her birth parents is her genes. In such a case, it makes sense that such a child would safeguard her sole inheritance as carefully as possible, would hold all gene tampering to an absolute minimum. Possibly the DNA, known possessor of great wisdom and skill, knows which of its vast assortment of genetic potential was actually expressed in the parents and in this way can follow the will of the child: "Take care! Modify as little as possible!" In this case, the child's free will architect wants nothing quite as much as to be ruled by the genes.

What of those people born with severe physical afflictions? Surely no one freely chooses this.

We all know that such severe afflictions do occur, so it must be a matter of either choice or chance.

If chance deals blows of such severity to people who have not chosen, and do not deserve, the afflictions they have, then free wheeling evil is most certainly loose in the world.

This may be, of course. Conventional wisdom for millennia has claimed that it is. But conventional wisdom has not always been proved right. We could eliminate one of the primary "proofs" of evil in the world if we could prove that people are not chance victims of any physical afflictions they may have, even those present at birth.

Proof of this is not yet forthcoming, as far as I'm aware, but an argument in its favor is not difficult to come by.

Is there any way to eliminate chance as the ruling factor when it comes to people born with severe physical afflictions?

For those millions of people who believe in reincarnation, explanation can be given very simply in one word: *Karma.*

In Hinduism and Buddhism, karma means *action,* specifically actions that bring upon a reincarnating spirit inevitable results, good or bad, either in this life or in a later life. In Theosophy, karma is the cosmic principle of rewards and punishments for actions performed in previous lives.

Throughout recorded history, far more people have accepted the theory of reincarnation than have rejected it. The belief is very old and has been held by primitive peoples throughout the world. Several

Greek sects believed in it, as did some ancient Jewish religious groups. Reportedly it was taught in early Christian circles. However, the Second Council of Constantinople condemned the doctrine in 553 CE and since then it has not been a part of Christian teachings.

Those who believe in the doctrine find that it helps make sense out of what otherwise seems an unfair, unjust world. Injustice becomes only apparent, not real. Karma providentially explains all the negatives away. Those who suffer do so due to poorly lived past lives. Those who live happily and successfully are reaping the reward for past lives well spent. It is neat and tidy, intellectually and emotionally satisfying, and, for those who accept and believe, it goes a long way to "vindicate the ways of God to man." Alexander Pope.

However, the fact that a majority of mankind has always embraced the doctrine of reincarnation, and that it makes beautiful sense to those who do embrace it, does not constitute proof of the doctrine's validity. Is there any way we can eliminate chance and elevate choice to a ruling position for those who view the notion of reincarnation as wrong-headed wishful thinking and reject it?

In the absence of karma, is there any other explanation as to why some human organisms might choose to be born with, or deserve to be born with, severe physical afflictions?

It is a generally accepted feeling, among those who reject continued existence on earth in reincarnated form, that we can and do achieve a certain amount of earthly immortality through our children and our children's children. We were instrumental in creating them and often see ourselves reflected in them. Even as we ourselves are heading upward toward better things in heaven, or are hell-bent toward the eternal fires of damnation…or are going wherever the dead go…some part of us still exists on earth in our children.

For the atheist or agnostic for whom death is often seen as leading nowhere, for whom the grave is often seen as the end of it all, here at least, in physical form, in our children, is the one clear shot we have at immortality. As long as our children and their children survive on this earth, a tiny part of us lingers on. We are not completely extinct.

That part of ourselves that we pass on physically to descendents is our genes, the long strands of DNA, the acknowledged master builder, the magical molecules with the blueprints.

Somewhere along the line can our genes decide that it is time to choose a severe physical affliction?

COMMENTARY

> *Sweet are the uses of adversity.*
> —— Shakespeare.

Possibly we can't quite bring ourselves to see the uses of adversity as "sweet," but it is easy to argue that adversity does have its uses.

Pain is nature's warning system and those incapable of feeling pain are at a grave disadvantage, are, in fact, in a truly perilous state. Pain shouts at us: "Danger! Switch course!" If we can't hear the warning, if we are deaf to pain, we can suffer critical burns before we're even aware we're being singed. In shouting its warning to us, pain can save us.

When an imperfect organism is created, have the genes decided that it is time to choose a severe physical affliction, with its attendant pain, as a warning signal: "Danger! Switch course!"?

As the DNA goes to work generation after generation, building one new organism after another, what if a tilt sets in, what if the organisms thus created are expressing themselves in a way that is faulty? Sooner or later wouldn't correction occur, karma not as applied to a reincarnating entity but action-reaction within the genes themselves? If a given trait has been duplicated over and over again, one generation after another, only to put the organism in physical or moral danger, surely in time mutation would occur, a sudden change to bring the problem to an end.

This is not to imply in any way that physically afflicted newborns are the offspring of people who in one way or another lack moral character. The moral sense, and the genetic pool that produces the beings who possess it, can vary markedly from one human being to another. Some people don't flinch at assaulting, robbing, killing members of their own species, while others brood if they accidentally snuff out a gnat. We can feel guilty over destructive actions committed or because we are striving mightily toward a perfection not yet achieved, but in either case the guilt is real.

How can we atone for our real or imagined transgressions? How does this powerful emotion express itself? If we can't face our own guilt and instead repress it, pushing it out of the verbal left brain into the creative right brain, what effect does this have on our genes?

Greater love hath no man than this, that a man lay down his life for his friends.

— John 15:13
Bible, King James Version

The newly conceived organism may, in love, choose to expiate the guilt splashing around in its genetic pool, may volunteer for the task of bringing a thorough cleansing to its genetic line. Sweet are the uses of adversity, and powerfully sweet must be a love ready, willing and able to take on and atone for the real and imagined sins of its forebears.

In seeking immortality by having children, we may be passing on to them more than we dream.

SUMMARY:

Envisionment: "The process or action by which the human will brings into material being its physical form, the human body."

If by any chance this is the way life works, we human beings are more godlike than we have yet dreamed we are, more completely in charge of our own fate. On the downside, if we are truly this powerful, then we must also accept full responsibility for the bodies in which we live. If we don't like the body we have, there is no one else to blame.

We can no longer view ourselves as victims and spend our life blaming others, or blaming fate. When we look in the mirror, we face the one who's responsible.

> *What is within us is also without.*
> *What is without is also within.*
> —— Katha Upanishads

If we wonder what our true feelings are, feelings buried so deeply within that we may not even know they're there, we have only to look in a mirror, study our faces and our forms, to find a roadmap of our hearts.

JUNK REVISITED:

According to the latest reports on the mapping of the genetic code, despite the overwhelming presence of "junk" material elsewhere in our DNA, within the densely packed gene cluster that lays out our anatomy, no "junk" is present. This would suggest there is some limit on our power to create. Apparently we can't decide, in the womb, to change our minds and instead of being born as a human baby come forth as a mouse or a whale instead. We are stuck with our original intent to be born to the human race.

8

The Demon Theory of Disease

❖

Modern Medicine: The Same Old Black Magic That We Know So Well?

> If I had my own way I'd make health catching instead of disease.
>
> —Robert Green Ingersoll

Throughout most of recorded history, magic and medicine have been inextricably intermixed. For primitive man, disease was an affliction caused by the entrance into the body of something evil, a force possessing the power to cause mental, emotional or bodily ills. How else rid the body of these unwelcome intruders than through magical rites and prayers?

> When the even was come, they brought unto him many that were possessed with devils and he cast out the spirits with his word, and healed all that were sick.
>
> —Matthew 8:16
> Bible, King James Version

Today so-called civilized man shares with Jesus and his contemporaries and with primitive man this same belief as to the cause of infectious disease. The names differ but the underlying principle remains the same. The medicine man calls the external forces that invade the body "evil spirits" or "devils." Modern, civilized man calls them "germs."

Where primitive man used only his instincts and his intuitions to tell him what caused disease, we have advanced far beyond this. We can now put our devils under magnification and watch them do their nefarious work. Or can we? As we watch our devils at work, what are we really seeing, or not seeing?

Modern medicine has several different ways to classify our wide range of physical disorders. One method currently in fashion separates them into ten major categories. For our purposes here, we need only distinguish between infectious diseases...those caused by demons...and non-infectious...not caused by demons.

A catchall term for our demons is germs, and we have quite as fancy an array of germs as primitive man has of evil spirits. There are disease producing bacteria, viruses, rickettsiae, mycoplasmas, and parasitic organisms such as certain amoebas and roundworms.

Each of these general types consists of many subtypes. Quite possibly we outdo primitive man...as well we should, with all our sophistication...in the number and variety of disease-producing devils we recognize and catalog.

These devils enter our bodies in ways that any child would be able to pinpoint. We can inhale the devil through our nose, ingest one with our food, or it can penetrate the skin or mucous membranes, most often through a break in one or the other. Once these devils have gained entry, they may or may not cause disease.

Sometimes they fail in their mission to harm us due to our bodily defenses against them. For those germs inhaled, we can sneeze or cough them out, or expel them with a runny nose. Against those vast numbers we take in with every bite of food, we wage constant chemical warfare. Saliva contains a powerful germ-killer that rids us of many of the enemy, and those who get by this first line of defense are most often destroyed by hydrochloric acid in the stomach. The third mode of entry, through the skin, is not easily accomplished, though our skin can be breached even where there is no open cut or wound in such places as the hair follicles or sweat glands. Some viruses...the most powerful devils we are up against...can achieve the near miraculous: They can penetrate even the unbroken skin or mucous membrane. Against such power, how can we possibly protect ourselves?

The fact that these devils often get past our defenses to set up residence within us still does not necessarily mean we will fall victim to disease. We can have what is termed a *sub-clinical infection.* Our health remains unaffected even though infectious agents have invaded our bodies.

Different germs cause illnesses through differing means. Some germs occupy cells and destroy them. Others produce toxins, or poisons, that make us ill. Each germ is also considered disease-specific, which is to say that a given germ produces a given disease and no other. Tubercle bacilli never cause anything but tuberculosis, the poliovirus causes polio and nothing else, etc.

Our modern demon theory of disease, as you can see from the above, is very logical and neat, but how valid...how scientific...is it?

For a branch of knowledge to be scientific, it must have a body of facts or truths systematically arranged which show the operation of general laws.

Our modern demon theory of disease is certainly "systematically arranged" but has medicine truly been able to show "the operation of general laws"? With all our high-powered electronic microscopy, when it comes to bodily ills have we really advanced significantly beyond primitive understanding?

How scientific...how removed from magical beliefs and practices...is our current medical thinking?

To try to answer this, let's first take a look at that minor scourge of us all, the common cold.

Modern medicine makes a distinction between infectious and contagious. The latter term usually refers only to those infectious diseases that can be spread directly from one human being to another.

Colds are considered both infectious and contagious. We "catch" them from one another. Your youngest child comes home from school feeling ill, and before you can say, "According to all the latest research, colds are caused by a virus, which is an infectious agent, any of a group of ultra-microscopic, infectious agents that reproduce only in living cells," everyone in the household is sneezing or coughing.

Or you go to work feeling just fine only to spend the day sitting next to someone who has watery eyes and a runny nose. A couple of mornings later you wake with a slightly sore throat and an immediate grudge. Darn that Clarabella for doing this to you!

On the other hand, of course, it may not work that way at all. Your husband may come home from work feeling rotten, moody with the wretchedness of an incipient cold. All family members rally round...physically as well as emotionally...hoping to cheer and comfort him. You feed him, kiss him good night, and sleep in the same bed. For a week he suffers the indignities attendant upon a cold while all family members sympathize...but no other family member gets the infection.

Or you have a luncheon date with a favorite cousin who shows up coughing and runny-nosed. Despite her cold, she enjoys seeing you and you spend several hours together, laughing, talking and reminiscing. The following morning you wake up feeling great, and the morning after that and the morning after that. It is five months before you catch your next cold.

How contagious is this particular "contagious" disease?

The common cold has been studied intensively by medical researchers. The name itself is something of a misnomer as there are at least two hundred viruses that cause cold symptoms. The most common infecting agent is the *rhinovirus*, "rhino" meaning nose. More than one hundred varieties of rhinovirus have been found, new strains appearing and disappearing with great frequency. Another family of cold-producing viruses are the *adenoviruses*. There are even some colds for which no virus can be found.

The accepted scenario for the full scale production of a cold runs something like this:

Someone with a cold sneezes or coughs, emitting into the air several million of his infectious germs. His hands, infected by contact with his eyes, nose or mouth, touch material objects around him, and upon these objects he leaves several million additional germs. A soon-to-be victim breathes in this virus-laden air, allowing invasion number one. As her hands touch objects previously handled by the cold sufferer, the germs swarming over the objects get onto her hands. Sometime later her hands touch her eyes or nose, leading to a second wave of rhinovirus invasion. She has by now been thoroughly contaminated with the highly infectious agent that caused the sufferer's cold and that will soon cause hers.

Cilia, microscopic hairs in the nostrils, propel the invading virus to the back of the nose where they occupy cells and begin reproducing additional viruses. In time, the viruses overflow the original cells, invade neighboring tissue and enter the bloodstream.

Meanwhile, your immune system, detecting the foreign presence, begins producing anti-inflammatory substances such as histamines and prostaglandin. As these substances fight off the invaders, they produce such symptoms as a scratchy throat, a cough and a runny nose.

The concentration of rhinovirus in the bloodstream begins to drop within two days of the onset of a cold, while at this point the symptoms themselves, caused by our fight against the virus, are growing more pronounced. In time, our bodily defenses figure out they have won, declare a cease fire, and we recover.

This is, in simplified form, the generally accepted scenario. If it's valid...scientific...then it must of course be reproducible under controlled conditions. Only when the same result can be obtained repeatedly from the same, controlled cause, can a theory be considered valid. With a sufficient number of successes, and with no failures, a theory eventually can be formulated as a "law."

Is there a "general law" underlying the above scenario as to how one person catches that highly contagious, infectious disease, the common cold, from another?

Right after the Second World War, in 1946, Christopher Howard Andrewes, an English doctor, launched an assault on the common cold. He became the first director of the Salisbury Common Cold Research Unit, housed in the Harvard Hospital and a large complex of buildings near Salisbury, England. Volunteers were lured to the facility in order for the staff to study the detection and transmission of colds. By 1970, more than seventeen thousand volunteers had visited Salisbury.

The volunteers were placed in isolation, two to a unit, and after several days, during which time physicians determined that none of them had colds, inoculations were given. Tests were double blind, so no one knew who had received cold viruses and who had been given instead a harmless saline solution. When the results were tabulated, it was found that the injection of cold viruses produced colds in 35% to 40% of the subjects injected.

35% to 40%!

With direct inoculation of the agent (rhinovirus) believed to cause colds, colds could be produced in less than half of all cases; 60% to 65% of the volunteers, after being injected with the agent believed to cause colds, failed to come down with any infection.

This is not only failure. This is failure on a pretty massive scale. Any theory that cold viruses, following successful entry into the body, cause colds should have been looked at with a jaundiced eye then and there. Viruses may be present when a cold occurs, they may be a contributing factor, but there is very clearly no direct cause-and-effect law at work when the failure rate is over 60%.

In another startling failure, the Salisbury volunteers failed to catch colds from one another. When volunteers with colds were paired with healthy volunteers, there was one cold in nineteen pairings at first. This was followed by one cold in five, then none in four. This is a remarkably poor showing for what is believed to be, and in the family unit is often experienced as being, a highly contagious, infectious disease.

Subjecting volunteers to chilly conditions, or having them wear cold, wet clothes for extended periods of time, also failed to produce colds.

To sum up, the Salisbury researchers seemed to have shown us this. When it comes to colds, you can infect some of the people some of the time, all of the people none of the time, and a few of the people never, a fortunate few who seem to have a built-in immunity.

In the mid-1970's, Dr. Jack M. Gwaltney, Jr., a professor at the University of Virginia Medical School, began his life's work, the study of colds. He successfully infected three student volunteers with rhinovirus. A few days later, when all three were coughing and sneezing, he gathered thirty-seven additional volunteers and divided them into three groups.

Those in group 1 spent three days in the same room with one of the infected students but weren't allowed to touch him.

Those in group 2 joined an infected volunteer around a table while the cold sufferer talked loudly, coughed, and sang.

Those in group 3 shook hands with an infected volunteer, then touched their own noses or eyes.

Of those who shared a room with a victim they never touched, none developed colds.

Only one of the twelve volunteers sitting at the table with a cold victim and subjected to prolonged coughing, singing and talk came down with a cold.

Eleven of the fifteen who had touched the cold victim fell ill.

This reinforced Dr. Gwaltney's suspicion that physical contact played a significant role in the transmission of colds. However, we should keep in mind that *four* volunteers...over 25% of the group...failed to get sick after experiencing what may be the easiest way for one person to infect another.

A few years ago researchers from the University of Pittsburgh conducted a study looking at the connection between colds and lifestyles. They found that "loners," people with few social relationships, were *four* times more likely to come down with colds than those who reported they had lots of friends. The researchers theorized that social interaction may strength the immune system.

A study from Wilkes University in Pennsylvania zeroed in on another way to strengthen the immune system: Frequent sexual activity. One hundred and eleven students reported on how often they enjoyed sex. Their levels of immunoglobulin A, an antibody that fights respiratory infection, were then measured, and it turned out that those students who reported having sex once or twice a week had significantly higher levels...one third higher...than those who had sex less frequently. But more of a good thing turned out not to improve the immune system even more, as the frequent-sex group had higher levels than the very-frequent group, the very-frequent group being defined as sex oftener than twice a week.

What hope is there for a cold vaccine to vanquish the common cold?

Currently researchers at the University of Ghent in Belgium are hoping to perfect a vaccine that will immunize us against all forms of flu. A protein has been found that seems to be present in every known flu strain, so a flu vaccine that affords protection against every form of flu may one day be developed.

If a similar protein could be found for cold viruses, there would be some hope of a vaccine to protect us from colds. But so far this is only a dream. With over two hundred known viruses all "causing" the upper respiratory distress known as the common cold, a protein present in all of them may never be found, and may not exist.

What scientific proof is there for the belief that viruses cause colds?

In the Salisbury trials, direct injection of rhinovirus into healthy volunteers failed to cause colds up to 65% of the time. When twelve volunteers sat around a table for Dr. Gwaltney and were subjected for hours to deliberately contaminated air, only one volunteer came down with a cold. Healthy volunteers who shared a room with a cold victim remained healthy.

Of the fifteen healthy volunteers who had direct physical contact with a cold victim, following which contact they touched their nose or eyes with their virus-contaminated hands, eleven caught colds. *But four out of fifteen didn't.*

If fifteen volunteers jumped off a two story building and, using only their arms and legs as an aid to flight, tried to stay in the air, and four of the fifteen managed to do this, we would be forced to go back to square one to figure out how gravity works. If the surest known way to transmit a cold from one person to another fails in four out of fifteen cases, possibly the theory that germs cause colds should be looked at more closely.

Having a lively social life apparently protects against colds, as does having an active sex life. Getting older helps. Those over sixty have fewer colds than younger people while children are the most frequent victims. It is possible to have no cold symptoms yet be not only infected but also contagious. Is some general law lurking somewhere in all of this?

In further pursuit of some general law substantiating the theory that germs cause disease, let's take a look next at the tubercle bacilli.

Tubercle bacillus is defined as the bacterium, *Mycobacterium tuberculosis*, which causes tuberculosis.

If someone infected with Mycobacterium tuberculosis coughs or sneezes into the air, anyone inhaling these infected droplets can develop tuberculosis. The primary stage of the disease is usually without symptoms and 95% of those infected will recover without any evidence of the disease.

Pulmonary TB develops in the small percentage of people whose immune systems do not deal successfully with the primary infection, or it may develop due to a second infection the body isn't able to resist. Tuberculosis may develop within weeks of the primary infection or can lie dormant for years before progressing into a disease. During these years when the Mycobacterium remains dormant, there is a sub-clinical infection in which health is not affected.

The fact that there is such a thing as a "sub-clinical infection" is positive proof…all the proof that any reasonable, open-minded person needs…that tubercle bacilli, *by themselves*, do not cause tuberculosis. Other factors must come into play inciting the dormant bacilli into a more active, disease-producing role.

The Los Angeles Times of February 19, 2001, reported the tragic death of two high school students from bacteria. One, a healthy senior, came down with what seemed like the flu. Suddenly, without warning, he died. A month later another student at the same school also died. In both cases, the cause was reported to be the same: A bacterial blood disease.

Later in the article it was reported that during the winter months about 20% of the general population harbors the bacteria "causing" this fatal blood disease, yet most people never fall ill…another "sub-clinical" infection.

We can harbor cold viruses...and even be contagious...without having colds. We can be infected for years with tubercle bacilli without developing TB. As the recent news article on the two tragic deaths indicates, we can play host to dangerous bacteria in our saliva and mucus without falling ill. If germs by themselves could cause disease, we'd all be in serious if not fatal trouble.

The medical profession explains the existence of sub-clinical infections by invoking the magic word *immunity*. If our immune systems are strong enough, we can fight off our devil-germs and not fall ill. It is when our immune systems fail us that the invading demons can do us in.

COMMENTARY

Currently there is a widespread belief that a positive attitude strengthens the immune system while a negative attitude weakens it. Our moods trigger the release of hormones, it is said, which increase or decrease our body's defenses.

This may be true, but it fails to answer one question that seems extremely relevant to me, which is:

> *Why at any given time in history are we immune to the ill effect of some germs but not to others?*

Tuberculosis has been with us since man became man. During the 19th Century it reached epidemic proportions in Europe and North America and was known as the "great white plague." As late as 1911, it was the leading cause of death in all age groups in the United States. Since then, in industrialized nations, the mortality rate has fallen dramatically. Currently in this country it is a rarity among otherwise healthy people. This is not the case in most areas of the world. Why are we immune when less fortunate people aren't?

Cold germs are everywhere, and not everyone who comes down with a cold has a negative attitude or a compromised immune system.

Optimistic people are stricken with cancer right along with their pessimistic sisters. In the recent report about the two high school students who fell ill and died, neither was reported to have had previous health problems. Why did their immune systems fail to protect them? Is there a hidden component in immunity not yet generally recognized?

HYPOTHESIS:

> *When a given disease expresses an emotional truth for us, we succumb. If it fails to express any emotional truth for us, our immune systems keep us safe.*

* * * * *

To attempt to prove there is validity in this, in the next chapter we will take a much closer look at colds to try to explain why they are still "common," followed by a discussion of smallpox, a once dreaded affliction that has now been conquered. Has it been vanquished because it no longer expresses any emotional truth for us?

9

Failure and Success: Colds and Smallpox

❖

Lose Some, Win Some...Is There Any Way to Better The Odds?

o o o o o o o o o o o o o o o o o o o o o o o o o o o o o
This is the noble truth of sorrow. Birth is sorrow, age is sorrow, disease is sorrow, death is sorrow.

—*The Pali Canon*

To begin exploring the hypothesis set forth...that the infectious diseases we succumb to express an emotional truth for us...let's start by taking an even closer look at that extremely familiar affliction, the common cold.

Different people experience colds in different ways. Some begin with a severe sore throat, then once the throat burning dies down, all other symptoms are mild or non-existent. Others may experience a mild sore throat, or no soreness at all in the throat, but become afflicted with what seems an endless nasal flow, a veritable upside down fountain of ill-smelling splash. Others may skip a heavily runny nose but suffer painful head and chest congestion. Some may get through the cold itself with only minor discomfort, then experience a hacking cough that lingers on for what seems like forever.

These are only some of the ways that we can experience what is known as the common cold.

There is not only a wide variance in the way cold sufferers suffer colds. The same person may experience colds in different ways at different times. Today a burning sore throat, six months ago an endlessly running nose, a year ago a bad night cough...each cold may highlight itself in a different fashion. Medical researchers tell us that over two hundred different "cold-causing" viruses have already been isolated. This may account for some of this variance...for those who accept the germ theory of disease...but any such germ theory leaves a great deal unexplained.

I may experience a cold largely with a wretchedly burning throat, then hand the cold over to my husband, and the same cold germ...presumably the same cold germ..."causes" in him a very mild sore throat but a painful head congestion. Why this different-type cold from the same cold germ?

Or I may experience each and every cold in precisely the same fashion, with a burning sore throat followed by a general recovery or extremely brief, very mild additional symptoms. If a certain type of cold-causing virus causes this type of cold experience, why am I resistant to all other cold germs and non-resistant to this one?

Once we are past the sore throat beginnings of a cold, we ordinarily spend the rest of the cold expelling and ejecting, ie, rejecting.

Rejection is almost always construed as an act of hostility, an act of non-love. We embrace what we love, reject what we fear, dislike, or for which we have no use. The body works within this general law. We evacuate our bowels, expelling matter that we don't need, for which we have no use. We urinate and perspire to rid ourselves of further waste. If we didn't rid ourselves of this waste material, we would soon die of our accumulated poisons.

A figurative expression of disdain...*I spit on you!*...tells us bluntly and succinctly what we are doing when we have colds: We are, quite literally, spitting on each other. However politely we may try to do this, hastily bringing a tissue to our nostrils when we sneeze, covering our mouths when we cough, we can't entirely disguise what we are up to: *I spit on you.* This is what a cold is, a compulsive spitting on the world, an expulsion of liquid and not so liquid phlegm from the body. Expulsion is rejection, ie, an act of hostility.

HYPOTHESIS:

> *The common cold is an expression of hostility, an expression of built-up anger or accumulated rage.*

COMMENTARY

If we have rage within us, it's a poison we need to get rid of. If the mind refuses the job, can't face its own anger, then the body takes over and comes down with a cold. Once infected, we can sneeze, spit and cough up anger by the bucketful. Once a thorough cleansing occurs, there is, most of the time, for most cold sufferers, a period of natural immunity before the buried resentments build up to discharge levels again.

Anger is a mental/emotional phenomenon. Therefore the body expresses it on the mental/emotional level, primarily through the nose, which is the emotional connection on the mental level. The degree of chest involvement depends upon whether or not the anger is superficial or deep.

What about the sore throat that so often precedes or accompanies a cold?

When we draw in air, the breath of life, the flow is separated in our nostrils, but then reunited in our throats. Our throats express an underlying oneness/unity.

If we are feeling so hostile that we invite a cold in to cleanse ourselves, this is a violation of the oneness/unity expressed by the throat. Such a violation shames us so we burn with it, and we burn with shame right where the violation is occurring, in the throat.

What of nose, head or chest congestion?
Those of us living in this age, the beginning of the New Millenium, are extremely familiar with congestion. Traffic congestion plagues our highways and byways, often reducing traffic flow to a crawl.

Head…nasal…chest congestion is a similar stoppage of flow. Where there is an over-abundance of accumulated anger, combined with a deep fear of expressing it…and if there weren't a deep fear of expressing it, it would never have piled up so abundantly…the flow of hostility is hindered. A bumper-to-bumper situation occurs, a bottling up that causes throbbing and aching.

Throbbing…"I want out! *No, you don't!*" "Yes, I do!" "*No, you don't!*"…the throbbing and ache of fearful indecision.

If the hostilities run so deep that they have been packed down, buried, stored out of sight and out of mind, congestion occurs in the chest, with a resultant coughing up of coagulated hostility: Phlegm.

With all these symptoms working in us and on us, we begin to feel we are dying and half wish we would.

But as the hours and days drag by, the poisons drain out, at their own rate of discharge, and soon we are feeling better, then in time recover completely, feel well again, thank goodness.

Yes, thank goodness, thank the good that is in us that triumphs, quickly enough, over the stored-up anger and fear. When suffering with a cold, we grow ashamed of our hostile feelings, cut off their expression and cure ourselves.

If we feel anger/fear, these negatives/non-usables/poisons must be expelled from the body, and colds are a reasonably safe way to express them. Where an open verbal expression of all our stored-up anger and resentment might bring numerous problems down on our heads, a cold ordinarily brings only physical distress.

Childhood is the most helpless, dependent, fearful age of all, so naturally enough it is the age most vulnerable to colds. What safer way to express the boiling anger within?

Loners...maladjusted adults with few friends and few social outlets...can harbor a great deal of anger that makes them more vulnerable to colds.

An alternative expression of anger...sex...if indulged in frequently can protect against colds, releasing anger in a more pleasurable way.

After age sixty, when the passionate fires of youth die down and we learn to accept who we are, what we have, and begin to truly appreciate what a blessing life is, we experience far fewer colds. We can in fact, as we lose our anger, become all but completely immune to them, experience an acquired immunity that has become natural for us. Colds become as much a part of our lost youth as adolescent acne.

Failing to recognize the cold for what it is, an expression of hostility, or in any case not taking this hostility personally, those around us most often offer sympathy. Loved ones, in many cases, hover near, taking warm, considerate care of us. We have an excuse to stay in bed, if this can be arranged, an excuse to be waited on and to feel and act grumpy.

Taken all in all, colds, no matter how miserable, are not all bad. Possibly this is why we have so successfully kept them in our lives despite all medical advances.

* * * * *

In direct contrast to the still prevalent cold, smallpox, a scourge for centuries, has been conquered. Is there some emotional truth this deadly disease expressed in earlier centuries that we no longer need to express?

It was well known in the English countryside, in the 18th Century, that milkmaids who became infected with cowpox from the infected cows they milked would thenceforth be immune to smallpox. In 1774 Benjamin Jesty, an English farmer, took cowpox matter from a pustule on the nipple of a cow and successfully inoculated his wife and two of his children against smallpox.

Dr. Edward Jenner (1749-1823), aware of the affinity between cowpox and smallpox, experimented first on his own son, Edward Jr., born in 1789. He injected the infant with the virus of pox from a pig. Eight days later the baby fell sick and several small pustules appeared. Sometime after the baby recovered, Jenner injected him with smallpox matter five or six more times without producing the least inflammation.

On May 14, 1796, Jenner scratched matter from cowpox sores into the arms of a healthy boy, James Phipps, who a week later came down with cowpox. After the boy recovered, Jenner injected him with smallpox from a pustule on the body of a smallpox patient. Nothing happened. The boy remained well. Jenner tried this a second time several months later. There was never another pustule. The boy clearly had immunity.

Dr. Jenner called the cowpox material "vaccine" and the method for its use "vaccination," both words being derived from the Latin *vaccinus,* meaning cow.

The use of Dr. Jenner's vaccination spread throughout Europe, then to America. In those areas where vaccination was known and used, smallpox was never again a serious epidemic disease.

This is truly a magnificent success story. Smallpox was at one time a scourge. Now, in those nations that mandate compulsory vaccination, it is a rarity. For this, we all owe a debt to Dr. Edward Jenner, also the milkmaid who drew his attention to the immunity from smallpox conferred upon those who'd had cowpox, and even more to the young boy, James Phipps, picked apparently at random by Dr. Jenner to be experimented upon. Had Dr. Jenner made some unfortunate error in calculations, the boy James, into whose body a virulent infection of smallpox was later introduced, might not have lived.

But the boy *did* live, and escaped smallpox entirely. On the basis of these two successes...his baby son and James Phipps...Dr. Jenner published his findings and in time became both famous and wealthy. Vaccination was on its way and the defeat of smallpox imminent.

Even prior to the above happenings, inoculation against the dread disease smallpox was not unknown in England. In 1721, Lady Mary Wortley Montagu returned to England from Turkey, where her husband Edward had been ambassador, and attempted to introduce into her home country the "heathen" practice she had learned from the Turks.

Every September while in Turkey, Lady Mary had watched a group of old women go from house to house in Constantinople, their purpose to practice preventive medicine. Each of the women would carry a small sample of pus collected from a victim of a mild case of smallpox. A vein would be scratched open on the limb of a customer, and the practitioner would dip her needle into the pus, smear it on the open vein, cover the wound and bind it. The process was known as "ingrafting."

Those subjected to this inoculation would remain healthy for eight days, then would suffer from a fever that might last for two or three days, following which they would recover and be as well as before their illness. Every year thousands underwent this treatment, there was no record that anyone ever died of it, and it offered protection against any future recurrence.

Lady Mary believed in the process fervently enough to have her own son treated, and in this way Edward Wortley Montagu became the first known Englishman to be inoculated against smallpox. He survived the inoculation and was thereafter protected against the dread disease.

Both the Turks' practice of "ingrafting" and Dr. Jenner's cowpox inoculation were based on the observed fact that people who have had certain diseases rarely get them a second time. Smallpox was one of these once-in-a-lifetime diseases. Those who survived were, in most cases, immune thereafter.

Modern medicine ascribes this immunity to the antibodies manufactured by the body in fighting off the initial attack. After the disease has run its course, the defense erected against it remains strong enough to ward off completely any subsequent assault. This is not an invariable rule, however, there being numerous exceptions to it. Dr. Jenner's successful vaccination showed that cowpox and smallpox were closely enough allied that the antibodies that remained following the cowpox attack would ward off smallpox.

Observation of the human condition has shown that there is not only the acquired immunity mentioned above, but that in addition a fortunate few have a natural immunity to certain diseases. When smallpox was devastating the Western world with recurring epidemics, on occasion as many as 75% of the population would be stricken, but even this horrendous percentage leaves 25% untouched. It is possible that a small percentage of this group suffered no exposure. Another percentage had surely acquired immunity from a prior onslaught. But there remain a fortunate few who, though exposed and not having any acquired immunity, nevertheless escaped the disease. Such fortunate people are thought to possess a "natural" immunity.

Where immunity is not total, there exists a variation in the degree of immunity. To put it another way, there are degrees to which different individuals are susceptible. Experience has shown that in any smallpox outbreak, the most severe cases occur first, those with the least immunity, the greatest susceptibility, being stricken not only first but hardest, followed by those with a somewhat greater immunity, whose experience of the disease will be less severe.

Along with varying degrees of immunity that can cause a lesser or greater severity of illness, the disease itself can appear in a milder or more severe form. It will be noted that in the Turkish inoculation program, the women medics who carried on the ingrafting program secured pus from victims of mild attacks. In 1789 Jenner described an outbreak of so mild a nature that no one died.

This milder version of smallpox has occurred in other parts of the world as well, and while some medical authorities consider it a closely allied but different disease, most classify it as smallpox no matter how mild.

The immunity to smallpox conferred by vaccination has been found to diminish with time, the amount of loss varying with the individual. In order to boost protection back to the highest level, re-vaccination has become accepted practice.

Another variable is this. Back when epidemics raged unabated, experience showed that the pestilence hit hardest where living conditions were crowded and unsanitary. Common sense would suggest that two factors may have been at work here. One, the more crowded the living conditions, the greater the exposure. Two, people living in slums may be less well nourished and generally somewhat less healthy, which could account for a greater susceptibility. Whether these two factors accounted for the difference, or whether other factors were also involved, is not known.

To sum up, we can make these statements about smallpox.

According to medical science, a virus causes smallpox...but only in the absence of natural or acquired immunity.

Smallpox is a dreadful, often fatal affliction...except when it appears in much milder form.

Those who have once had smallpox are immune thereafter...except in a few rare cases.

Vaccination affords protection...but the level of protection decreases with time, the amount of lost protection varying with the individual.

Uncrowded, sanity living conditions afford some protection…but far from total immunity.

In these statements about smallpox, we come up with quite a number of exceptions and variables. Even the basic statement, "A virus causes smallpox," has to be qualified, for the virus does *not* cause smallpox in those who have a natural or acquired immunity.

There is a startling difference between an established law of physics and those beliefs that pass for "law" in the field of medicine. The United States could not possibly have put a man on the moon if our scientists had been dealing with laws of motion as uncertain, subject to as many exceptions and variables, as this "law" that smallpox, a serious often fatal disease, is caused by a virus.

The hypothesis was offered in the previous chapter that emotional factors are at work in the production of disease, that the diseases we succumb to express emotional truth for us.

Can we pinpoint any emotional truth that smallpox may have expressed for centuries that we no longer feel any need to express?

To attempt to answer this, in the next chapter we will take a much closer look at the once feared plague.

10

Smallpox: A Closer Look

♦

Did Dr. Jenner Have Help in Conquering the Pox?

o o o o o o o o o o o o o o o o o o o o o o o o o o o o o o o
Adversity is the first path to truth.

―*Alfred Lord Byron*

L et's take a close look at what happens when the body comes down with smallpox.

A typical case manifests itself about like this:

Prior to the onset of symptoms there is an incubation period, believed to be from ten to fourteen days, then the disease strikes suddenly and severely. First there is a pronounced chill, soon followed by a fever. The temperature rises to 104 degrees Fahrenheit or even higher, while the sufferer experiences a quick pulse, intense headache, vomiting, and pain in the loins and back.

These symptoms continue for three days, during the course of which time rash resembling scarlet fever or measles may appear on the lower part of the abdomen and inner sides of the thighs.

On the third or fourth day following onset, the characteristic eruptions begin to appear, almost always on the face first, most often about the forehead and roots of the hair. The eruption spreads over the face, trunk and extremities, always most marked on the exposed parts of the body. On the second or third day, the original eruption, the *papules* (a small, somewhat pointed elevation of the skin, usually inflammatory but non suppurative, ie, without pus) change into *vesicles* (a skin elevation containing serous fluid). These vesicles show a slight central depression.

The clear contents of these vesicles slowly become turbid, and by the eighth or ninth day are changed into *pustules* (a small skin elevation containing pus). These fully developed pocks grow larger and lose the central depression.

There may be noticeable inflammation and swelling of the skin. Eruptions may also be present on the mucous membranes, and the eyes may also be involved. With the appearance of the initial eruption, the temperature quickly drops, but in many cases there is a return of the fever as the vesicles are converted into pustules.

This fever may be accompanied by great restlessness, delirium or coma. A few days later the pustules begin drying up and the fever drops. The skin begins itching. The dried pustules produce scabs that gradually fall off and reddish brown spots remain, which leave permanent white depressed scars. The "pitting" so characteristic of smallpox is especially noticeable on the face. If the patient survives, the probability is great that he will be immune to further attacks.

Let's take a close look at this appalling array of symptoms, translating them...as best we can...into their emotional equivalents.

The onset of smallpox is sudden and severe, first rigor, soon followed by fever.

Rigor: a sudden coldness, as that preceding certain fevers.

Our language tells us what emotion we are caught in when we are cold. We can feel "cold with fear," or "frozen with terror." Coldness can imply death...*rigor mortis*, the stiffening/coldness of death... unconsciousness, ie, "out cold"...or sexual frigidity. To have "cold feet" is to lose courage, to be afraid. To be a "cold fish" is to be one who displays no love, passion or affection.

We never say of anyone that he is "cold and loving" anymore than we would describe someone as "warm and unloving." Warmth goes with loving, coldness with being unloving, and with being afraid.

Coldness/fear can be such a devastating emotion that to function at all we must push it away, and the energy with which we push it away is anger. If we are frightened enough, we reach for the energy of anger to save us. If we are angry, underlying the anger is a fear we haven't yet faced.

Smallpox starts with fear, a fear that is soon pushed away by anger, ie, a fever. Fever is an elevation of temperature, hotness.

Our language tells us what feeling hot means: "Hot tempered," "Hot under the collar," "Hot and bothered"...all expressions linking being hot with feeling angry. There is another recurrent linkage as well, expressed picturesquely in the phrase "Hot to trot."

"Hot" often carries a connotation suggesting physical excitement tending toward sexual excitement. We can become aroused...worked up...heated up...either in anger or with sexual passion, each being an alternative expression of the other. A "fever" is an expression of hotness/anger/sexual passion.

Smallpox starts with fear, but the fear is soon pushed away by anger/sexual passion, the energetic emotion that saves us from being "frozen with fear," that saves us from total paralysis.

While in the grip of fever, the smallpox victim experiences a quick pulse, intense headache, vomiting and pain in the loins and back, also in some cases a rash on the lower part of the abdomen and inner sides of the thighs.

A quick pulse: When we are afraid or angry our hearts beat fast.

An intense headache: A headache feels as though ignorant armies are clashing on and in and through our brains, which quite possibly is what a headache is: A clash/battle in the mind.

With the onset of smallpox, the body is being forced to carry an extremely heavy burden and may not survive. Possibly the mind should open up and find some other way to deal with the negatives rather than pushing them off on the body. But the airy, avoidance-prone mind would often rather die than face certain unpleasant truths about itself, so the mind darts away while the body, staggering under the horrendous load of negative messages registering on it, tries to avoid having to express them all, and the battle is on. The head hurts in keeping with the hell of a war being waged within it.

Vomiting: Food is the primary physical expression of love. When we vomit, we are rejecting food, nourishment, ie, love.

Pain in the loins: Anger is the other face of fear. Sex is an alternative expression of anger. The anger the fever is expressing could be channeled into sex, and the loins ache with the hunger to do this. Any pain says: "*Pay attention! Remember me!*"

The pain in the loins says: "Hey, you guys, you up there in my head fighting away, you who are so busy stirring up my blood to fever pitch, remember me. I'm your loins and can handle this for you, if only you'll let me." But this message, if it ever gets through, is quickly suppressed...*for the fierce war going on is being fought to keep the loins out of action, not to give in to their base desires.*

Pain in the back: Man is distinguished from all other animals by the spine that keeps him upright. The pain in the back says: "*Pay attention: I'm an upright guy!*" Man is unique not only because he stands upright on two feet, but also because he has a year-round, not a periodically recurring, sex life. The upright stance...the spine...was instrumental in bringing this about and pain in the back should bring this to mind. The back pain reinforces the message of the loins: "*We could deal with this fever through sexual expression if only you dumb, frightened guys would pay attention!*"

There may occasionally be a rash resembling scarlet fever or measles on the lower part of the abdomen and inner sides of the thighs during the pre-eruptive stage of the disease. Here the body is getting right down to cases, setting up a flashing red signal right around the accident scene, sending out a screaming reminder of how this entire problem could be handled...through the genitals...but who would dare pay attention to this hell-fire-and-brimstone solution?

Fortunately, after three or four days of this, the eruptions begin to pop out and the fever dies down. The heat is off...for the time being at least...as the conflict has been resolved. The body is now expressing the accumulated anger/sexual passion in a non-sexual way.

HYPOTHESIS:

A skin eruption equals a "non-sexual" sexual release.

In smallpox the eruptions are first papules, a small, somewhat pointed elevation of the skin, usually inflammatory (reddish, swollen, warm, tender) but without the production or discharge of pus. Then they change to vesicles (containing serous fluid), then with another change they become pustules (containing pus) at which time the smallpox victim may experience another fever, accompanied by great restlessness, delirium or coma.

Compare this to the experience of sexual release:

With the first sexual arousal in the male, there is a "small somewhat pointed elevation" of the penis, usually inflammatory (reddish, swollen, warm, tender) but not suppurative (no pus). With continued arousal, a clear, slippery "serous" (lubricating) fluid appears. With activity carried to climax the penis ejaculates, ie, ejects semen, a process often accompanied by great restlessness (movement), delirium (of joy) or coma (some people lose consciousness at orgasm. In earlier ages, "to die" was a euphemism for experiencing orgasm).

Pus: a yellow-white, more or less viscid substance found in sores or abscesses.

Semen: a whitish, viscid fluid produced by the male reproductive organ.

* * * * *

If there is the least substance in the contention that an eruption on the skin equates with the erection of the penis in sexual arousal, we will surely find corroboration for this apart from the pox of smallpox. Let's take a quick look at acne.

Acne: an inflammatory disease of the sebaceous glands, characterized by skin eruptions, especially on the face.

Acne tends to be a plague of adolescence, blighting those teenage years between puberty and adulthood. The body matures sexually at puberty to where it becomes capable of reproducing itself (the definition of puberty), which in common law is considered to be fourteen years in the male, twelve in the female. Oftentimes, while the body may be ready at these early years, the thinking/feeling human being inside the body still has a ways to go, and sexual frustration may occur. During these years of sexual frustration, the pus-filled skin eruptions of acne often occur.

Years ago I was delighted to hear Orson Bean, when asked on a TV game show whether he still remembered his first lover, affectionately mention a girl's name and then remark with gratitude that she had saved him from spending a fortune on acne medications. When I mentioned this to a Latina friend, she laughed and said that it is commonly accepted folklore among the Mexican people that acne is an expression of sexual frustration. Adult males tease pimpled teenage boys, she said, that all they need is a good woman to clear their skins.

HYPOTHESIS:

> *Smallpox can be considered a dangerous, sometimes lethal form of acne, or acne a mild form of smallpox. Both are poxes upon our*

physical houses, an expression of sexual frustration displayed on the skin.

* * * * *

But if there is any validity in this, why would anyone in his right mind choose the misery, death-defying risk and almost sure disfiguration of smallpox over the generally pleasant experience of sexual intercourse?

Possibly because people are ruled by many different emotions. Strong beliefs may determine our choices. Many early Christians chose martyrdom, which at first glance may seem irrational. If we look closely enough, we may find that the choice of smallpox over sexual expression in some contexts makes excellent sense.

If one sincerely believes that life on earth is brief and matters only as a preparation for eternal bliss, it is not too difficult to understand why such a sincere believer would cling to his faith, and to his virtue, even in the face of death.

Smallpox raged as a plague during those centuries when the Christian Church, with its message of the sinfulness of sex and the glories of eternal life that awaited the righteous, reigned supreme.

For a sample of the Church's attitude on sex, consider the essay *Of the Honorableness of the Marriage Bed* by St. Francis de Sales (1567-1622), there being, of course, no such thing as "honorable sex" apart from the "marriage bed."

> *The Marriage Bed must be "undefiled"...that is to say, exempt from immodesty and other defilements. Thus was marriage first instituted in the Earthly paradise where, until the time of the fall, there was no disorder of concupiscence in it, nor anything dishonorable.*
>
> *There is a certain resemblance between the pleasures of the flesh and those of eating, for both of them relate to the flesh, although the former, by reason of their brutish vehemence, are alone called*

carnal. I will therefore explain what I cannot say of the former, by what I shall say of the latter.

St. Francis, in discussing the "honorableness" of marital sex, has already cast several severe aspersions upon it. Marriage as first instituted, ie, prior to the time sex reared up in all its ugliness, was obviously to be preferred, for then it had nothing "dishonorable" in it. Concupiscence is a "disorder," the pleasures of the flesh have a "brutish vehemence." With this kind of defense, what is left for any who might wish to launch an attack?

St. Francis continues by pointing out that just as food is necessary for the preservation of life, sex is necessary for the procreation of children. Therefore it is "good, holy and commanded." One spouse is not to withhold sex from the other, but is to allow it faithfully, freely, with good spirit…as long as sex is not indulged in "to excess and immoderately…"

> *Of a truth, nuptial intercourse which is so holy, so just, so commendable, so useful in the commonwealth, is nevertheless in some cases dangerous to those who make use of it; for sometimes it makes their souls very sick with venial sin, as happens by simple excess; and sometimes it causes them to die of mortal sin, as happens when the order established for the procreation of children is violated and perverted, in which case, such sins are always mortal…*

Continuing with the analogy he has set up…eating as similar fleshly pleasure standing in for sexual indulgence…St. Francis writes:

> *It is a true mark of a beggarly, mean, abject and base spirit to think of the dishes and of eating before the time of the repast, and more so still, when afterwards one is taken up with the pleasure which one has had in the meal, dwelling upon it in words and thoughts, and allowing one's mind to wallow in the remembrance of the pleasure enjoyed in swallowing down the mouthfuls; as do those who before dinner have their mind fastened on the spit, and*

after dinner on the dishes, persons fit to be scullions...Persons of honor do not think of the table until they sit down to it, and afterwards they wash their hands and mouth, in order to lose both the taste and the smell of what they have eaten.

St. Francis then explains how the "honorable" discord of sex should be handled:

The elephant is only a huge animal, but he is the most worthy beast that lives on the earth, and the most intelligent. I will give you an instance of his excellence; he never changes his mate and tenderly loves the one of his choice, with whom nevertheless he mates but every third year, and then for five days only and so secretly that he has never been seen to do so, but he is seen again on the sixth day, on which, before doing anything else, he goes straight to some river, wherein he bathes his whole body, for he has no wish to return to the herd until he has purified himself. Are not these excellent and modest traits in such a beast, by which he invites married persons not to allow their affections to remain attached to the pleasures of sense which they have experienced in accordance with their state of life, but, when these are past, to wash their heart and affection of them, and to purify themselves of them as soon as possible, that afterwards they may perform other actions which are more spiritual and lofty?

This sets it up very clearly for the devoutly believing Christian:

If one strives for excellence, sex should be indulged in but once every three years. One should purify oneself immediately afterwards, never think about it beforehand nor allow the mind to enjoy the memory of it afterwards. Anything beyond this is excess and excess is dangerous. It will sicken the soul. Sex was "established for the procreation of children" and if this natural "order" is "violated or perverted," it is always a mortal sin and the sinner will be dammed.

Who could be sure of living up to these strictures, which covered not only the arena of action but extended into the mind as well? Who could keep all stray thoughts of sex away, could keep himself from anticipating beforehand or enjoying in memory afterwards? Yet such lapses were venial sins that sickened the soul. To go even further, to dare to pervert sex, brought nothing less than eternal damnation. To lose one's life was a relatively minor thing, but to burn forever in the fires of hell, forever and ever, time without end...No wonder the pox looked almost benign in comparison.

It is interesting to speculate on whether, or how often, the plague flared up coincident upon the emergence of some eloquent new preacher with a fiery tongue able to preach mighty sermons of warning to the eager faithful, who trembled in their minds and hearts as they listened and crawled home chastened...possibly to be stricken within days with the chills and fever that signaled the pox.

It is not suggested that only sinners were smitten with smallpox, or that the worse the sinner the worse the pox. The reverse was more likely true: The more delicate the conscience, the more piercing the sense of sin, the greater the susceptibility to smallpox. After all, it all depended on what one feared. All the pox could do was kill you, while being disorderly, dishonorable or perverted in your sexual thoughts or actions could damn you to an everlasting hell. Little wonder that in those sin-ridden centuries, the plague flourished as it did.

With the industrial revolution and the age of enlightenment leading into the modern age, smallpox was effectively conquered, along with...for most of the Western World...the deep, dark, all pervading sense of sexual sin.

To further explore the meaning of, or purpose behind, disease, in the next chapter we will look at another once-dreaded infectious disease, yellow fever.

11

Yellow Fever

◆

The Color Disease: Lemon Yellow Skin with Black Vomit

> I never wonder to see men wicked, but I often wonder to see them not ashamed.
>
> —*Jonathan Swift*

Yellow Fever, the most dreaded of all plagues in the Americas, first traveled to the New World on slave ships in the 17th Century. Victims turned lemon yellow and then the terrifying black vomit started. The Latin American name for the fever was *el vomito negro*...the black vomit.

Epidemics swept through America's port cities. Between 1702 and 1800, the deadly fever struck thirty-five times. It is estimated that half a million Americans contracted the fever, and one hundred thousand of them died. In 1793 the fever swept into Philadelphia and within two months carried off 10% of the population. After several months the plague lifted and survivors observed that the arrival of colder weather had seemed to kill it.

During the 19th Century, despite all the precautions taken, the epidemics worsened. The fever struck ports from Boston to New Orleans and was rife throughout the Caribbean.

The fever was a powerful ally of the indigenous people of the Caribbean in their struggle against colonization. The British, French, and Spanish lost whole armies to yellow fever. In Haiti the fever routed Napoleon's troops, and in Panama, it, along with malaria, forced the French to abandon construction of their canal.

During the Spanish-American War, yellow fever raged among the American troops in Cuba. When the war ended, a commission headed by Major Walter Reed was dispatched to Cuba to study the plague. It arrived in the summer of 1900.

For some years a physician in Havana, Dr. Carlos J. Finlay, had been claiming that mosquitoes transported yellow fever, but he had not been able to convince anyone of this. When the four-man American commission headed by Major Reed, after intensive study of eighteen yellow fever victims, failed to find any cause for *el vomito negro,* they went to call on Dr. Finlay to listen to his outlandish notion about mosquitoes.

Dr. Finlay turned over to the commission some small, cigar-shaped black eggs and told them they now had the eggs causing the fever.

Jesse Lazear, one of the commission members, carefully stored the eggs in a warm place so they could hatch into wigglers. In time the wigglers grew into extremely pretty mosquitoes, with silver markings on their backs.

To prove that mosquitoes carried yellow fever, experimental animals were necessary. At that time it was erroneously believed that no animal was susceptible to the disease except for man. For the mosquito theory of transmission to be tested, man would have to serve as the experimental animal.

But how could anyone deliberately give human beings yellow fever? In some epidemics, records indicated that eighty-five out of a hundred victims died of it. In other outbreaks, fifty out of every hundred died. In the mildest outbreaks, at least twenty out of a hundred died. To deliberately infect a human being was tantamount to murder.

But if nothing were done, epidemics would continue to rage, endangering American soldiers. The decision was made to go ahead, using volunteers.

Lazear took his pretty, silver-striped, female mosquitoes, and walked down between the rows of beds on which lay men sick with the fever, allowing the mosquitoes to bite and suck their fill.

These same mosquitoes, once they'd digested their meal of infected blood, were applied to the limbs of eight volunteers and allowed again to bite and suck. But, alas for the theory of mosquito transmission, all eight volunteers stayed healthy, with no sign of yellow fever in any of them.

Major Reed had been ordered back to Washington to report on their work. Dr. James Carroll, another member of the commission, mindful of Major Reed's orders to test out the mosquito theory, insisted that they try again. He had Lazear bring out the most dangerous mosquito in his collection and had himself bitten by it. Four days later he came down with the fever, the first victim of deliberately induced yellow fever.

When Dr. Carroll showed signs of the fever, the same mosquito that had bitten him was set up to bite another volunteer, an American soldier, William Dean, and he too was stricken.

At this point ten men had been experimentally bitten, and two of them had fallen ill.

It was later determined that the first eight men had failed to succumb not due to any natural immunity but because the mosquitoes used to bite them had bitten the yellow fever victims too late in the course of their disease. Although the mosquitoes were gorged with yellow fever blood, it was not infectious blood.

When Dr. Carroll and the soldier William Dean were successfully given the fever, the insect causing their illness had been hatched in the laboratory and had fed twelve days before upon a yellow fever patient, who was then in the second day of the disease. This combination of factors brought about their successful inoculations.

Dr. Carroll's case was quite severe, so much so that a fatal result was expected for several days. After recovering, Carroll became convinced that the severity of the disease depended upon the susceptibility of the individual and not on the number of bites he had gotten. To test this out, on October 9, 1901 he had a volunteer bitten by eight contaminated mosquitoes. The attack that resulted was mild.

Once Dr. Carroll and William Dean had been successfully inoculated, further controlled tests were necessary to prove that the mosquito bite and it alone had caused their disease.

New volunteers were isolated, kept under isolation long enough to make sure they weren't already infected, then bitten by mosquitoes who had gorged on infected blood. Other volunteers were put into small houses, fourteen feet by twenty feet, into which were brought bedding and pillows soiled with black vomit. Every possible source of contagion, other than the mosquito, was tossed in, with later volunteers even sleeping in pajamas worn by men dead of the fever.

The results were that when the mosquito was allowed in to bite, yellow fever resulted even under the cleanest, most sanitary conditions. But as long as the mosquito was barred, it did not matter what other elements were present...how soiled and foul smelling everything was...no yellow fever resulted.

But what if the men sleeping on the pillows and sheets, and under the blankets, and in the pajamas of men who had died weren't susceptible? Maybe every single one of them was naturally immune.

To eliminate this possibility, two of the volunteers who had inhabited the stench-filled little house were subjected to a further test. One had yellow fever blood shot under the skin, the other was bitten by infected mosquitoes. Following this, both were stricken, so no natural immunity had protected them earlier. In these kinds of controlled tests, Dr. Carlos Finlay was proved right. His little black eggs were indeed the cause of yellow fever. Mosquitoes carried yellow fever from one victim to the next.

Once this had been established, there was an all-out campaign to rid Cuba of the *Aedes aegypti,* the highly domesticated carrier mosquito. Less than a year after the intensive clean-up campaign was undertaken, Havana was free of *el vomito negro* for the first time in one hundred and fifty years. After similar campaigns, yellow fever disappeared from the United States. By 1925 it had been banished from Mexico and Central America and a clean-up campaign had been launched in South America.

This is a beautiful, soul-stirring success story if ever there was one. Paul de Kruif wrote triumphantly in 1926 (in his book *Microbe Hunters*) that there was barely enough yellow fever poison left on earth to put on the points of six pins.

While de Kruif's jubilation was natural enough, his statement was, unfortunately, somewhat premature. By 1928 yellow fever had returned in some parts of the world. A vaccine developed against it was used for mass vaccinations occasionally launched by the World Health Organization to deal with outbreaks.

In time it was discovered that yellow fever, like smallpox, is "caused" by a virus.

There are other similarities to smallpox. Once is usually enough, one attack assuring most survivors a lifetime immunity. The degree of susceptibility varied also. During the course of the controlled studies in Havana, two volunteers, a soldier named Kissenger and a civilian clerk named John J. Moran, were put into preparatory quarantine and later bitten by infected mosquitoes. Kissenger became ill but Moran did not.

Yet that Moran had only a partial "natural immunity" was proven later when he underwent additional, repeated bitings and in time did succumb. Of five Spanish immigrants who agreed, for the sum of $200 each, to "volunteer" for the experiment, four were stricken after being bitten while one was not.

One striking feature of the Cuban trials was this: The recovery rate for the commission's volunteers was phenomenal. The first two to be successfully inoculated, Dr. Carroll and William Dean, both recovered. The next three volunteers all recovered. John J. Moran, once the mosquitoes finally got to him, recovered.

This recovery rate was with a disease that often claimed 50% to 85% of its victims.

The only early casualty was Jesse Lazear. Although Lazear was a member of the commission and one of the first to volunteer to be bitten, his death did not result from a planned inoculation. After the initial failure in which he participated, he became skeptical of the theory that mosquitoes transmitted *el vomito negro* and in consequence didn't bother to brush away a mosquito that had settled on the back of his hand as he visited a ward filled with yellow fever victims. A few days later he came down with the fever and six days later he died.

As with smallpox, yellow fever could produce a mild illness, a severe one, or one that was fatal.

With yellow fever there is natural immunity...acquired immunity...varying degrees of susceptibility...just as in smallpox or other infectious diseases.

Let's take a closer look at what all of this might mean.

Yellow fever arrived in the New World on slave ships. It ravaged port cities where the slave trade flourished. Victims turned lemon yellow and their vomit was black.

In the epidemic of 1793, Dr. Benjamin Rush, the most eminent physician in Philadelphia, attended patients from dawn to dusk. This is his testimony as to the facial appearance of his patients (as quoted by Carol Eron in her marvelous book, *The Virus That Ate Cannibals*):

> *It was as much unlike that which is exhibited in the common bilious fever as the face of a wild is unlike the face of a mild domestic animal. The eyes were sad, watery, and so inflamed in some cases*

as to resemble two balls of fire. The face was suffused with blood, or of a dusky color, and the whole countenance was downcast and clouded... sighing attended in almost every case.

HYPOTHESIS:

Yellow fever, like smallpox or the common cold, expresses a specific guilt/anger/fear, in this case guilt over slavery and fear for what it might mean to a slave-owner in the hereafter.

* * * * *

Consider the observations Dr. Rush made about his stricken patients:

Sighing. We often sigh when we feel depressed or deeply regretful.

Eyes sad and watery, ie, tearful. Tears can be caused by grief, regret, fear, or by positive emotions such as joy, but here the eyes are "sad" as well as tearful, which would rule out joy. The eyes are also in some cases inflamed with blood, ie, anger/passion. Accompanied by sadness, this red blood/anger/passion would seem to be turned inward upon the self, not outward upon the world.

Face suffused with blood. We can be red-faced with either anger or shame, but in this case, with the "whole countenance...downcast and clouded," the face would seem to be blood-red more in shame than in anger.

Facial Appearance: The look of the face suggested to the good Dr. Rush the difference between the face of a "wild" as distinguished from a "mild domestic animal."

COMMENTARY

Yellow fever journeyed west with the slave ships. On these ships were black men and women believed to be "wild," uncivilized, not quite human compared to the civilized, "mild domestic" white men and women who were enslaving them.

Yellow fever ravaged the eastern and southern coast, that is, all the port cities. All or most of these cities were active in the slave trade.

Our founding fathers wrote in the Declaration of Independence that "all men are created equal," but this use of the term "men" did not include women, who were not considered equal, nor did it include those with black skins, who were also considered inferior.

To be female or black was not only to be something less than equal. It was also, at least in the case of the blacks, to be considered something less than human. As late as 1900 Christian preachers in this country were still giving sermons to "prove" that black people "have no souls." (Earlier in history such sermons had been given to "prove" that women "have no souls.")

This denigration of the black race into something less than equal, something not quite human, had to be accepted by slave traders and slave owners, for how else could one be both a good Christian and a dealer/trader/owner of slaves?

But what if...*what if*...the suspicion began to creep in that slaves weren't "wild" animal creatures after all, but were instead fully human, people who had every right to be treated as equals, with respect and dignity? How did a slave-owning population deal with that, especially in the port cities where the importation and sale of slaves flourished right under its nose?

If this horrifying awareness of the humanity of black people couldn't break through into the mind, it could break through to color the skin, color it with the yellow of awareness...and cowardice...causing the eyes to become sad and watery, also in some cases inflamed with anger directed inward. It could suffuse the face with shame, a face that began to look like that of a wild animal. When white people were enslaving blacks, who was really the beast? Then there was the terrifying vomit, vomit that was black, black to show that inside the white skin lived the same blackness, the same humanity...or inhumanity...that belonged to the slave. Under the skin, all were alike, all were brothers, just as Christ had taught.

If black people brought in chains from their home continent into slavery in a faraway country were indeed human and equal, then the white Christians enslaving them had better beware of the peril they were in, the eternal damnation they faced if they didn't open their eyes to face the enormity of what they were doing.

But to become aware...oh, sweet Jesus, who had repeatedly commanded us to love one another, to love our neighbors as ourselves...how could the mind accept such a staggering guilt? Let the body carry the burden, express the awareness and the pain, even if the fever proved fatal. Better that than to have the mind crack under the strain of such an insupportable weight.

Those in Cuba who volunteered to be infected with yellow fever in order to help in its eradication had an astonishing recovery rate. One can assume that they felt far less crippled by guilt than those who fell victim to the disease under less humanitarian circumstances. Therefore their disease would reasonably prove to be milder and non-lethal.

With a general recognition by the white Christian world that the enslavement of their black brothers was an evil not be condoned ever again, yellow fever was conquered throughout that part of the world which no longer countenanced slavery. It has since reappeared on occasion, however, in other areas.

* * * * *

After yellow fever reappeared and had become once again a subject of intensive research, it was decided to try to culture in the laboratory a strain known as the Asibi strain, which was considered the most lethal yellow fever virus known.

And where had this "vicious" strain come from?

From a twenty-eight-year old black man named Asibi, who lived in a village on the Gold Coast of Africa.

And what of the poor man who had harbored a virus of such lethal intensity?

He felt ill for a time, then he recovered and went back to work.

The doctors studied the lethal virus strain they had found in Asibi. Possibly they would have learned more had they studied the man with such incredible resistance.

Would that we all had the disease-fighting potential of Asibi.
Do we have the potential to become as disease-resistant as he was?
This is a question I'd like to explore in the chapters that follow.

12

Health, Good and Otherwise

◆

If We Prefer the Good, Why Do We So Often Succumb to the Bad?

○ ○
It is impossible to live pleasurably without living prudently, and honorably, and justly; or to live prudently, and honorably, and justly, without living pleasurably.

―――*Epicurus*

When we come to consciousness and find ourselves in a human body, how can we best deal with the predicament this poses? What is the best way to assure ourselves of comfort as we work within our physical limitations, the best way to avoid the long list of ills the flesh, blood and bone are heir to? In brief, how can we maximize our pleasures and minimize our pains?

We know that some people are fortunate enough to have a natural immunity to certain infectious diseases. Others have little susceptibility. They have to be repeatedly exposed before the infection takes.

When it comes to non-infectious afflictions, ranging all the way from a minor headache to terminal cancer, the degree of resistance also varies widely. Some people can claim in all truth that they haven't been sick a day in their lives. Others spend half their lives in bed or dragging themselves painfully out of bed. A blessed few live what seem to be entirely pain-free lives, are rarely troubled even by a minor ache. Others keep enough medication on hand to stock a pharmacy, pop pills on the slightest provocation, yet remain pain-afflicted even when heavily drugged.

How can we align ourselves with the pain-free...if that is our wish...and avoid the anxiety and discomfort of disease?

Is there anything approaching a sure-fire way to avoid disease, to avoid being ill, sick or ailing?

For starters, we can of course pay attention to what the doctors and dieticians tell us: Eat properly, don't overdo, get plenty of rest, learn how to relax and avoid undue stress. Don't smoke. Drink alcohol only in moderation. Exercise regularly. Maintain a recommended body weight. Cultivate an optimistic attitude. Have regular medical check-ups. With this regimen, if you're lucky, you'll feel reasonably well and will stay fit.

But what if you aren't lucky? What if, in spite of your best efforts to follow all of the above, your body seems to have it in for you and malfunctions frequently out of what seems to be nothing but plain orneriness or spite?

We can all look around and see those who break many or all of the above-mentioned health rules yet somehow, for some unfathomable reason, get away with it, while others adhere religiously to them and rarely experience the joyful vigor of bursting good health. Why? And how can we align ourselves with the former and break rank with the latter? Are there some important health rules not generally known that, if only we followed them, could assure us our best shot at staying happily fit?

As we have seen in prior chapters, germs do not "cause" disease, at least not all on their lonesome. We can harbor any number of dangerous and even deadly germs without falling ill. Something else is clearly at play besides those old devil germs.

Medical science calls the other player in the game *resistance* or *susceptibility*. The higher our resistance, the less our susceptibility. We can have partial or, if our resistance is high enough, total immunity to certain germs.

How can we best strengthen our resistance and lessen our susceptibility?

COMMENTARY

Possibly we can start by adopting as a working hypothesis the theory argued herein that the diseases we succumb to express an emotional truth for us.

Throughout our history, human beings have most often been viewed as composed of three aspects, body, mind and spirit or soul. If we give the matter any thought, we will soon realize that we are not our bodies nor are we our minds. Rather, something within us...some inner sense of self...possesses a body and possesses a mind, but in itself is not either one. It matters not whether we call that which possesses without being possessed the soul, the self, the spirit, or something more fanciful. In an earlier chapter I referred to it as the architect.

Throughout the centuries many have believed that we human beings, possessing both minds and bodies, are here on a spiritual journey. Pierre Teilhard De Chardin, French philosopher and priest, reversing this in a wonderful turn of phrase, said of us that we are not human beings on a spiritual journey but spiritual beings on a human journey.

However that may be, one of the things that happens to us on our journey is that most of us on occasion fall ill.

If our illnesses express emotional truth for us, who is in charge of deciding that we will express our truths that way?

HYPOTHESIS:

Germs, no matter how powerful, cannot cause any infection on their own. While they play a part in the production of disease, they

act only when they have been ordered to do so. We order up a disease in the same way we order up a steak, because it suits us.
Who is handing down the order?
The same architect (the human will or spirit) who oversees construction of the body stays on the job to supervise all necessary maintenance, including renovations and, when unavoidable, destruction.
We all know that repairs...renovations...cause discomfort and inconvenience. Nevertheless, they are meant to serve a purpose and invariably do, in the human body quite as much as in any other edifice.
Some force within the body wills the body's diseases, their onslaught, their severity, their duration, and whether or not recovery will occur. When it comes to our diseases, we hold the reins of power, not the invading germs.

The human body is constructed in such a way that it can survive only when it receives frequent nourishment. For the body to be nourished, two conditions must be met. There must be an intake of food, and the body must utilize the food taken in. A superabundance of food will not keep a body alive and healthy if the mechanisms within the body to process and utilize the food have broken down.

In like fashion, when it comes to infectious diseases, two conditions must be met. There must be an invasion of germs, and the human spirit must put these germs to work to create disease.

Disease, like bodily nourishment, is cooperatively produced. If there's no cooperation, there's no disease.

* * * * *

But why would the architect...the human spirit or will...ever okay the infliction of a serious or deadly disease upon the body it occupies?

It we are in fact on a spiritual journey, as millions have believed throughout our history...or, even more telling, are spiritual beings on a human journey, as de Chardin wrote...it becomes fairly easy to answer the question as to why our spirits would agree to visit disease upon our bodies. If we must learn a certain lesson and stubbornly refuse to do so, the pain of disease may prove to be the only language we are open to hearing.

Consider this situation:

A painter enters a room intent on painting the air. She sets up her spray gun, fills it with yellow paint, aims it at the air around her, pulls the trigger and sprays. The air becomes cloudy and yellowish, but every time the painter turns off the spray, the air clears rather quickly again. It does not stay painted.

With every new attempt, however, the spray gun splatters additional paint onto every solid object around. Soon the floor, ceiling, and walls are bright yellow, as is all the furniture in the room. The painter has still not achieved what she set out to do. She has not successfully painted the air. Instead she has accomplished what she did not set out to do. She has painted the room.

In like fashion, the spirit within us may try interminably to get a certain message to break through into our conscious minds, but the airy mind floats away on the slightest breeze, refusing to listen. Meanwhile the physical body hears the insistent message and pitches in trying to help. In this way, we become the "victims" of disease.

Consider this scenario. You are due to run in a marathon, twenty-six plus grueling miles. For years you have trained as a runner, you love running, you want to run in this race. But at the last minute a problem arises.

You put on newly purchased running togs and nothing fits. The sweatband around your head is so tight it will guarantee a headache. Your shirt is so tight you cannot move freely, your shorts even tighter.

Worst of all, your running shoes are at least three sizes too small and will all but cripple you.

In your overly tight clothes it hurts you to breathe, your limbs are constricted, and your feet...oh, sweet mother, there is no way you can run in these shoes! But the race is about to begin, the gun goes off, you have no choice. Off you go, in your ill-fitting clothes, struggling forward as best you can.

Or let's say the problem is of exact opposite dimensions. You dress for your marathon only to find that your newly purchased clothes are several sizes too large. The sweatband meant to stay snugly in place on your brow is so loose it refuses to stay where it belongs, instead keeps dropping down over your eyes. Your shirt is so many sizes too large that it will flap wildly in the breeze, the shoulder seams being down below your elbows. Your pants are so loose they pose an ever-present danger of falling down. But worst of all are the shoes, gunboats so long and wide that even tightly curled toes may not succeed in keeping them on.

Oh, dear mother, there is no way you can run in these clothes, in these shoes! But the race is about to begin, the gun goes off, you have no choice. So off you go in your ill-fitting clothes, struggling forward as best you can.

Fanciful? Yes. No marathoner would ever allow herself or himself to be caught in such an fooish predicament. But when it comes to going forward in life, *this is the predicament most of us are in most of the time.* We outfit ourselves in ways that do not fit.

To return to our imagined marathon, let's give no further thought to our benighted runner and consider instead what is happening to the ill-fitting clothes.

The too-tight clothing, subjected to the runner's attempts to loosen the constriction it imposes, is going to be under constant strain. If the runner rises above her/his pain and keeps running, the overly tight clothing will gradually loosen…until it bursts a seam or two. In time it will tear or split to where it becomes unusable as clothing unless/until repaired. Damage has put an end to its proper functioning.

The too-loose clothing will also be subjected to immediate, excessive wear and tear. Flapping sleeves catch all too easily on protruding surfaces, which may tear them. If droopy pants fall, they can be dragged along a rough surface and shredded. Shoes too large to keep on, stepped out of and abandoned, have no way to protect themselves against being squashed down by others or run over. Soon all of this clothing too will become unusable without repair. Damage has put an end to its proper functioning.

COMMENTARY

Our primary predicament as human beings is that we outfit ourselves improperly. All too often nothing we have taken on fits us properly and we don't know how to make adjustments to bring about a proper fit.

A proper marathon can be run only when the clothing worn fits the runner. The clothes worn and the body that wears them both belong to the physical world.

The spirit in each of us possesses a mind, and no matter how tied to the physical body both spirit and mind may be, neither is expressed in the physical world in the way that clothes and the physical body are. No one has ever seen a mind, or a spirit. Scientists cannot put either one under a microscope. Both spirit and mind are non-physical.

Based on the above, I'd like to offer some suggestions on how good health can be maintained.

GOOD HEALTH HYPOTHESIS ONE:

> *Just as, on the physical plane, for ease of movement and healthy, happy functioning, the clothes must fit the body, on the non-physical plane the mind must fit the spirit.*

Good health exists when the mind is a proper fit for the spirit. The better the fit, the more radiantly joyful the health.

In life, most of us run a much longer race than a twenty-six mile marathon. During the course of the race...as in any long race...the power in charge, the spirit, is forced to make increasingly stringent demands. The mind must keep up with these demands or the body, feeling the incessant pressure, will feel the strain. With prolonged strain, the body cracks and damage occurs, just as it does to the ill-fitting clothes of our imagined runner.

GOOD HEALTH HYPOTHESIS TWO:

The mind stops being a proper fit for the spirit when it becomes polluted with too many unexpressed negative feelings.
Just as physical waste material must be expelled from the body or we'll become ill, possibly even die of our own accumulated poisons, so must negative emotion be dealt with and discharged. If there is a build-up of repressed fear, anger, guilt or resentment to where the spirit, offended by all this poison, feels pained, the spirit will demand a cleansing. If the mind refuses to handle this cleansing, the body is forced to take on the job, discharging the negatives through a process we call disease.

GOOD HEALTH HYPOTHESSIS THREE:

Bodily strain, with possible resultant damage, occurs precisely where the indwelling spirit experiences lack of a proper fit.
If our shoes are too tight, our feet hurt. Our hurting feet let us know exactly where the tightness is, where the strain is.
In identical fashion, any physical discomfort we experience pinpoints the area in which the spirit feels pain.

* * * * *

We have looked closely at three infectious diseases to see how this works. Smallpox would make no sense today here in the United States. In these post-sexual revolution days, unmarried couples talk freely of their sexual lives together on national TV without the slightest hint of a blush, or, as far as one can tell, the least worry as to what such behavior will do to their immortal souls. It makes perfect sense that smallpox has been conquered.

If yellow fever still raged, it would make no sense. As a country we have banished legal slavery and though we clearly still have a long ways to go to achieve a color-blind society, we don't suffer the shame anymore of dehumanizing an entire race. It makes sense that in the United States, we no longer are attacked by recurrent visits of *el vomito negro,* the black vomit.

That we still haven't conquered the common cold also makes sense. Until we become saints, we are apt to nourish some anger and to need a way to discharge it. It's hard to imagine a simpler and less dangerous way than through the common cold.

In the following chapters, I will take a close look at some non-infectious conditions or diseases to see if they fall into any observable pattern, and whether or not they reflect the same kind of etiology, or cause.

13

Vision Problems

❖

See Spot Run. See Dick and Jane Squinting to See Spot Run

○ ○
Where there is no vision a people perish.

——*Ralph Waldo Emerson*

In preceding chapters, we have dealt with infectious diseases. Now let's take a look at a typical non-infectious condition.

Just as the common cold is probably the most widely experienced infectious disease, there is a physical disorder so commonly experienced in our midst today that almost everyone has some acquaintance with it: The need for eyeglasses to aid sight. Though we ourselves may have escaped this plague, we almost certainly know someone who is nearsighted, farsighted, or suffers from astigmatism, eye fatigue or some other condition which requires the periodic or continual use of glasses to boost or clarify vision.

Let's take a close look at an extremely common vision problem: Myopia or near-sightedness.

Those afflicted with this condition can see clearly close up, but objects at a distance are blurred. For objects to be seen clearly the image must be focused on the retina, which is at the back of the eye. In myopia, objects at a distance focus at some point in front of the retina.

There are several explanations offered for this error of refraction.

One explanation offered...the one most widely believed...is that the eye is too long, from front to back, causing light rays from distant objects to focus short of the retina. Or an eye of normal size can become myopic, it is believed by some, due to an increased curvature of the cornea and lens, causing too great a refraction...bending of light...resulting in the same shortened focus.

The most widely held view is that myopia is an inherited condition. Those holding this view claim that there is no evidence that over-use of the eyes, even in poor light, will influence myopia, or that it can be helped by doing eye exercises or by wearing glasses.

Another school of thought believes that myopia is caused not by the inherited shape of the eye but is brought on by eyestrain, from too much reading or other close work done under too poor a light.

Nutritionists advance the theory that myopia is a disease of the connective tissue of the eye. In 1956 it was found that the fibers of collagen in the eye of a myopic individual have a smaller diameter than the fibers found in a normal eye and that they are bathed in a more abundant liquid. Due to these two factors, the collagen fibers of the eye may become abnormally distended and unable to give proper support, so that the ligaments become lax.

Another researcher, Dr. W. H. Bates, an ophthalmologist who studied the problem of myopia and other vision problems through long years of experiment and clinical practice, concluded that near-sightedness is the result of strain leading to poor visual habits.

The mode of dealing with myopia varies with the belief as to its cause. Most commonly, instead of pills to pop to correct the condition, glasses with concave lenses are prescribed. Just as most pills today provide relief from the symptoms without effecting any cure, eyeglasses provide normal vision but are not a treatment. A crutch, yes...eyeglasses or contact lenses...but not a cure.

Those who ascribe to the nutritional theory of myopia claim that the widely accepted view outlined above is in error and that it is possible to halt the progress of the condition. Medical researchers tried prescribing vitamin E and claimed that they succeeded in stopping the progress of the disease so that their young patients did not become more nearsighted.

Checking eight years later, they found the effects were lasting and declared that the cause of myopia is lack of vitamin E, either because there is an exaggerated need for it or because it isn't properly assimilated. Not only did they achieve a complete stoppage of the myopia itself. They also claimed there was improvement in some general symptoms associated with it, joint disorders, glandular upsets and retarded puberty.

Dr. W. H. Bates, who considered myopia as due to strain and poor visual habits, claimed repeated cures with his method of relaxation techniques and visual reeducation. Visual accommodation...the ability of the eye to focus first on a near object, then on one farther away...is accomplished, Dr. Bates contended, by the external muscles of the eyeball. These muscles are functioning improperly if myopia is experienced. So why do these muscles so often fail to function properly?

Dr. Bates claimed to have demonstrated in thousands of cases that this abnormal functioning of the external muscles was caused by a strain or effort to see. Once this straining to see was corrected, the action of the muscles became normal and all errors of refraction disappeared. As long as the eye doesn't *try* to see, the external muscles act normally and normal vision results.

In this myopic age, many of us have preconceived notions as to what type of person wears glasses. If we were asked to visualize first a college professor and then a street mugger, we would quite likely put glasses on the former and not on the latter. Various studies have shown that such stereotypical thinking, quite possibly derived from experience, has more than a little validity. Myopes do tend to share certain traits that distinguish them from non-myopes.

Psychological testing of one hundred and forty male myopes and one hundred and eighteen male non-myopes belonging to the entering class at a military academy showed that the myopes scored higher on academic achievement and creative performance while the non-myopes were more oriented toward business, selling skills and outdoor activities. The myopes were significantly more likely to achieve officer status and win more academic awards, according to an article by F. A. Young, R. M. Singer and D. Foster, "Psychological Differentiation of Male Myopes and Nonmyopes" published in 1976.

In another study, Raymond M. Singer of Washington State University, researched a possible relationship between visual refractive error and various personality factors and academic performance. In earlier studies, Mr. Singer noted, myopes had been described as introverted, more interested in ideas than in action, and academically oriented.

For his own research, aimed at delineating any relationship between psychological variables and myopia, Mr. Singer worked with approximately 300 male and female university students. The test subjects were examined visually and given psychological surveys. Grade point averages and pre-college aptitude test scores were obtained. The results?

Myopes scored higher on almost every measure of academic aptitude. They earned higher college grade point averages. They tended to prefer mental to physical activity and liked to work independently. They valued wisdom and independence more than their non-myopic peers and valued being clean and helpful less.

Myopes scored lower on the degree of super-ego (in Freudian terms, the conscience) and on proneness to guilt. They tended to respond more rationally to life situations than do non-myopes. They cared less about living an honest or exciting life, and more about living an intellectual life than did their non-myopic peers.

Mr. Singer concluded that it is possible to identify individuals who will develop myopia before the condition appears so that they can be referred to eye doctors for preventive treatment.

To suggest that once pre-myopes are identified they should be referred to eye doctors for preventive treatment seems a rather curious conclusion to draw from the research results that precede it.

Ophthalmologists...eye physicians...generally accept the idea that myopia is an inherited condition, that properly fitted glasses can correct the myope's vision but can do nothing to cure the condition or halt its progressive nature.

Even if all future myopes could be identified through psychological testing at age four, a couple of years before myopia usually makes itself evident, what could the eye physicians possibly do to help? All they can do is prescribe glasses, and to put the pre-myopic child in glasses before they are needed would be like putting someone on crutches before and not after she breaks her leg.

In Mr. Singer's study, it was found that there is a *psychological* difference between myopes and non-myopes. Therefore if you classify individuals prior to the onset of myopia on the basis of their psychological characteristics, in order to send them for preventive treatment, surely it is to a psychologist that they should be sent.

According to Mr. Singer's study, if you could subject the pre-myopes to psychological conditioning to where they would learn to value wisdom and independence less, value being clean and helpful more, could also inject them with greater conscience and a higher level of guilt, teach them to put less value on the intellect and on mature love...if all this could be accomplished, the myopes would no longer differ from the non-myopes and possibly through this psychological reconditioning the onslaught of myopia could be averted.

Also if we could catch children young enough and make sure that none become introverted, that all grow up to be more interested in action than in the world of ideas, and above all make sure that none become good students who are oriented toward academic achievement, we might possibly wipe out the myopic eye for good and all.

Another study on *Personality Factors Related to Development of Myopia*, in which subjects were given various personality tests, found that moderately near-sighted people were found to be the most different from those with normal vision, and that these moderately myopic people saw the world as noxious, irrational and pressing in on them.

This intriguing finding tends to suggest that myopes blind themselves to the world *out there* because they find it unbearably ugly and too close for comfort. The treatment suggested in connection with this study was relaxation and guided fantasy.

There is a school of thought that maintains that the myopic eye is a failure of evolution. For eons man, the hunter, needed distant vision, not near vision, and his eye, adjusted for distant vision at rest, perfectly suited his needs. But with the advent of mass literacy, newspapers, books, and the by-now ubiquitous computer screen, the demands made upon the eye for near-seeing have simply overwhelmed it. The unnatural strain of forcing accommodation to all these near objects has been more than the eye could handle, and widespread myopia is the result.

We can have our modern print culture and poor eyesight, or we can return to a more primitive way of life and once more enjoy the normal vision still the fashion among less civilized people. In this view, normal vision has been sacrificed upon the altar of civilization and the myopic eye is the eye of the future, for all of us.

Dr. W. H. Bates, who spent his adult life studying visual problems, considered this nonsense. The creative force that fashioned us did not stick us with a defective, non-adaptive eye. Primitive people did and do a great deal of near work. The women are seamstresses, embroiderers, weavers and artists who create a great deal of fine and beautiful work.

Yet women living under primitive conditions have eyesight that is as fine as that of the men.

It is not because modern, civilized man does a great deal of near work that he loses normal vision, according to Dr. Bates, but because, straining to see, he falls into poor visual habits and loses the natural art of seeing. Relaxation, reeducation and eye exercises will correct all errors of refraction, Dr. Bates insists.

He found the myopic eye to be "greedy." Too anxious to see everything all at once, it has lost the art of shifting the way a good eye should. With normal vision, objects are visually recorded portion by portion a bit at a time. The myopic eye is in too big a hurry for this and attempts to take in the entire forest all at once, unmindful of the individual trees.

Myopes also tend to stare too intently, dimming their vision as a result. In addition, the myopic eye is too dry. It fails to blink and tear sufficiently. The eye with normal vision sees one part of what it looks at best, and all other parts less clearly. This is what Dr. Bates calls "central fixation." The myopic eye, operating under strain, can no longer see best the point at which it is looking. It has lost the relaxed art of seeing through central fixation.

Once the myope has fallen into these bad visual habits, she compounds the problem by grabbing immediately for crutches. She gets fitted for glasses. This puts an immediate end to any hopes for improvement and ensures that her vision will deteriorate even more.

COMMENTARY

With vision problems, as with all other breakdowns in bodily functioning, the medium is the message.

The eye exists entirely on the mental level and connects directly with the brain.

According to Dr. Bates' long, intensive study, the myopic eye has lost central fixation, is greedy, stares too intently, and is too dry, failing to blink and tear enough.

Central Fixation: The central feature of anyone's life is his own being, primarily his feeling self for above all else we are creatures dependent upon our hearts. Myopes have lost this natural, easy fixation on their own emotions. This is verified by studies showing that myopes are primarily interested in the world of ideas, that academically they are high achievers, that the myopes' approach to life is intellectual, not emotional.

Greediness and Staring: The myope's intellectual approach to life includes the drive to take everything in at once, to stare down the world, to see life steadily and to see it whole. How else can one achieve understanding and control? And without understanding and control, how can one hope to attain wisdom and independence, traits highly valued by the myope?

Failure to blink and tear: To blink is to lose sight of, and when the world is seen not as a beautiful, restful place but as noxious and irrational, who would dare close her eyes, even momentarily, and lose sight of the noxious irrationality pressing in on her? Eyes that tear are eyes that cry, and to cry is to be emotional and centrally fixated in a way the myope has forsworn.

When we are walking in the dark with the aid of a flashlight, we illuminate the path at our feet or those other nearby areas that might threaten our safe forward progress. When we drive our automobiles at night, we turn on our headlights, not to see what is miles away but to check out that which is immediately in front of us. Often when we attend a stage presentation, a spotlight directs our attention to the performer we are meant to watch, while everyone else on stage fades away.

These visual aids...flashlights, car lights, spotlights...though lighting up narrowly restricted areas, nevertheless give us the help we need. With these aids, we are in much the same situation as the myope, who can see clearly and distinctly close-up but for whom distant objects are fuzzy and blurred.

HYPOTHESIS:

The message of the myopic eye is this: If you want to see life steadily and see it whole, if you want to understand it, control it, become wise and gain independence, forget what's "out there," at a distance, which is of little or no importance to you. Study that which is near to hand, look inward, know yourself.

Only because this message is not sufficiently absorbed by the conscious mind, and is insufficiently acted upon, does the eye become nearsighted to start with and ever more nearsighted as the guiding spotlight contracts into a closer and closer range, doing its best to drum in the message:

If you would know the world, know it steadily and know it whole, first of all know yourself.

The myopic eye can relax into normality, can regain central fixation and normal distance vision, once this message gets through and is acted upon, just as the flashlight and car lights and spotlights can be switched off once such visual aids are no longer needed.

* * * * *

Let's take a quick look at another visual problem, one that's the reverse of myopia: Hypermetropia or hyperopia, farsightedness.

Whereas approximately one-fifth of the adult population now experiences some degree of myopia, hyperopia is even more widespread, is in fact the most common of all problems of refraction. Estimates are that it affects up to two out of three of all adults.

Like myopia, it is believed by the medical profession to be caused by the shape of the eye. In hyperopia the eye is too short, from front to back, or, far more rarely, the curvature of the cornea is insufficient. In either case, the image picked up by the eye is not focused properly on the retina, but comes to a focus behind the retina. Convex lenses correct the error of refraction.

Whereas the myope rarely to never experiences any symptom of eye strain...distant objects are fuzzy but there is no problem in viewing them...those afflicted by the opposite condition, farsightedness, commonly experience headaches and eyestrain when they are forced to use near vision.

COMMENTARY

Here again, as with myopia...as with any bodily malfunction...the medium is the message. The clear distance vision is a spotlight directing the attention of those afflicted.

HYPOTHESIS:

> *The message of hyperopia is that it causes strain and pain to be so wrapped up in the near, so stop paying so much attention to every petty little object and happening close to hand. Divorce yourself from the near. Gain distance and perspective. Stop concentrating on every bug on every little tree and learn to view the forest.*

Here again, as with the myopic eye, it is only because this message is not sufficiently absorbed by the conscious mind and is insufficiently acted upon, that the farsighted eye becomes and remains farsighted. Once a sufficiently distant, detached view of the world is achieved, so that every little happening no longer causes strain and pain, and once this message is acted upon, the hyperopic eye can relax into normal close vision, just as any visual aid can be switched off the moment it's no longer needed.

* * * * *

If as a people we Americans don't achieve vision, we will perish, according to the man still regarded by many as the most inspirational writer in American literature, Ralph Waldo Emerson.

Possibly we should ponder this thought before we rush in every few years to alleviate vision problems by ordering new glasses. When do we start paying attention and stop suffering from increased loss of sight?

Leaving eye problems behind, in the next chapter let's take a look at another non-infectious condition, another extremely common affliction, one that causes an immeasurable amount of pain and anguish: Arthritis.

14

Arthritis

❖

*The Thigh Bone Is Connected To The Hip Bone,
The Hip Bone Is Connected To The Backbone...*

o o o o o o o o o o o o o o o o o o o o o o o o o o o o o

Pleasure is oft a visitant, but pain
Clings cruelly to us.

—*John Keats*

The term "arthritis" refers to an affliction, sometimes inflammatory, of one or more joints. There are more than one hundred different kinds, with many different causes. Around forty million Americans...one in six adults...suffer from this disease in one form or another. One family in three in this country is affected.

The disease is older than man, and has plagued mankind for at least one hundred thousand years. The skeleton of a Neanderthal man has been found with the spine almost intact...a spine deformed by osteoarthritis.

With this long history of familiarity with the disease and its widespread occurrence, how much do we now know about the cause, treatment and cure?

Not very much.

Let's look first at rheumatoid arthritis, "the great crippler."
Rheumatoid arthritis is a disease of chronic, fluctuating inflammation of one or more joints. Up to three million Americans suffer from it, three-quarters of them women. Symptoms most often appear between the ages of twenty-five and fifty. It may begin in only one joint and progress to others. Fingers, hands, wrists, knees, ankles, shoulders, neck and hips are commonly affected, but any movable joint can become arthritic.

Symptoms tend to fluctuate in severity. Joint pain and swelling and morning stiffness are common. Both sides of the body are affected. There may be irritability and flu-like symptoms, also fatigue and weight loss. It is considered a systemic disease in that it affects the entire system.

The onset of the disease can be gradual or abrupt. In up to 20% of those stricken, it may, within ten to fifteen years, disappear completely. Remission is most likely to occur early on and the probability of it decreases with time. After fifteen or twenty years, only 10% of victims will be severely disabled. Average life expectancy is shortened by three to seven years for most victims, while those who are totally disabled may die ten to fifteen years earlier than their non-arthritic peers.

The cause of rheumatoid arthritis is not yet known, but currently it is believed to be an auto-immune disease. Ordinarily the body can distinguish between its own cells and foreign invaders, but in auto-immune disease, this distinction gets lost. In rheumatoid arthritis, antibodies generated by the immune cells go on the attack against elements of the connective tissue in joints as they would against foreign substances.

There are three suspected causes for this auto-immune response: Microbes, outlaw immune cells, and leaky gut syndrome.

Increasingly, rheumatologists support the germ theory, chief suspects being viruses, streptococci, the tuberculosis bacillus, or...the most favored...microbes called *Mycoplasma organisms.* However, new findings from research teams suggest that immune cells may take guidance from other, defective immune cells...outlaw cells...that may trigger misguided, sustained attacks on targeted tissues.

The third suspected cause, a leaky gut, occurs when spaces between cells in the intestinal wall become enlarged. This allows large molecules to pass into the bloodstream, where antibodies treat them as foreign substances subject to attack. Finding a similarity between these large molecules and molecules in joint tissues, they mount an attack against the joint tissue as well.

Other underlying causes or contributing factors may be diet, nutrition, toxic drugs or emotional stress.

There is no cure for rheumatoid arthritis. It usually requires lifelong treatment aimed at relieving the symptoms.

Treatment consists of a many-sided strategy that may include disease-modifying and anti-inflammatory drugs, suggested exercises, rest, physical aids such as canes, splints or crutches, or the application of heat or cold. Advice may also be given on how to cope by adopting a new lifestyle.

Another approach is nutritional. Rheumatoid arthritis patients, it is claimed, have been found to be deficient in certain vitamins and minerals. Some of these nutrients are known to help bolster the immune system and to minimize inflammation. It is suggested that those suffering from RA should take not only a regular multivitamin tablet but additional supplements, as needed, of vitamins A, B, C, D and E, and the minerals boron, copper, zinc and selenium.

In the past, nutritionists have claimed significant success treating rheumatoid arthritis through dietary means. In one experiment, twelve hospitalized arthritis patients were put on a raw food diet, with no other treatment given. Eight began to feel better, two improved for a time before relapsing, while two showed no improvement at all. In the follow-up, after the patients who improved had gone home, it was found that seven out of eight continued to improve.

Vitamin E therapy has been shown to be beneficial. In a recent German study, it was found that taking 1,200 mg of vitamin E a day reduced the pain, stiffness and tenderness of rheumatoid arthritis every bit as effectively as taking a non-steroidal anti-inflammatory drug. Danish research over twenty years suggests that Vitamin E helps prevent RA.

Researchers studying the relationship between diet and arthritis claim that mounting evidence shows a strong connection. Certain vitamin deficiencies, it is claimed, influence certain types of arthritis. Eating a healthy diet and maintaining a normal weight, besides being good for overall health, help to avoid or manage arthritis.

A recent study of five patients indicates they were helped by what it is hoped may prove to be a breakthrough therapy, one that removes B lymphocyte cells from the bloodstream. These cells produce the antibodies believed to cause RA. Further studies are necessary to substantiate the results.

If all else fails in alleviating the ravages of RA, surgery may be suggested. The damaged old parts are yanked out to be replaced by new. This is usually thought of as a last resort when pain, stiffness, disability, and deformity are such that the sufferer can no longer cope with her life.

Surgery is not considered a "cure" but a confession of failure, an admission that medical science has yet to solve the mystery of the disease. At this point in time RA cannot be prevented, adequately controlled or cured. The treatment of last resort...surgery...offers relief for a particular joint, that is all. The underlying arthritis is unaffected.

In the past various studies were done on the psychological characteristics of people who suffer from RA. In one study, as reported by Harold Geist, Ph.D., in his book, *The Psychological Aspects of Rheumatoid Arthritis,* published in 1966, the researchers found that those suffering from arthritis shared certain personality traits. They appeared to be calm and rarely expressed any hostility. They did voice complaints about being shy and feeling socially inadequate. Any anger they expressed appeared to be directed at themselves.

Another investigator suggested that arthritis may begin with aggressive impulses that are unacceptable to the person feeling them. She erects a defense of muscle tension against these impulses and then represses the conflict. In this theory, inhibited aggression is the underlying cause of an essentially self-inflicted illness.

Other researchers have also considered aggressive impulses to be a prominent part of the etiology of arthritis. Arthritics are described as having a great deal of repressed hostility, angers and resentments that are primitive in nature, destructive and sadistic. These feelings are far too frightening to face or to allow expression. This results in emotional self-restrictions and an inability to feel or show anger.

It was also found that RA sufferers responded remarkably well to placebos. Whereas with most diseases the effectiveness of placebos…sugar pills passed off as medications…averages about 35%, with RA sufferers the response has been up to 60%.

Such research findings and suggestions were not universally approved of or accepted. Susan Sontag in her book *Illness as Metaphor,* published in 1977, a book which deals not with arthritis but with tuberculosis and cancer, vigorously condemns any such approach. Sontag labels such views, which make the patient responsible for his own disease, as dangerous and preposterous.

Such thinking apparently won the day and research into possible links between psychological traits and RA either died away or went underground. Dr. Arthur Freese in his book *Help For Your Arthritis and Rheumatism,* subtitled, *All the facts your doctor doesn't have time to tell you,* published in 1978, in discussing the treatment of RA does not at any time mention any need to deal with inhibited rage.

The same is true for the book *The Arthritis Bible,* published in 1999, written by Craig Weatherby and Leonid Gordin, MD, in which no mention is made of inhibited rage as a possible causative factor. However, these authors do recommend certain psychological approaches, such as cognitive/behavioral therapy, guided imagery and relaxation techniques, which they say have been proven effective in helping those suffering with RA to deal with their pain.

A study reported in the Journal of the AMA in 1999 offered proof of a mind-body link in the ravages of RA. Participants in the study kept diaries. Those who wrote of stressful life experiences were rewarded with a reduction in their symptoms, while those who wrote of emotionally neutral topics received no benefit.

Another recent study, reported in the April 2001 issue of the *Journal of Pain*, indicated that patients who have strong religious or spiritual beliefs can use this to cope with their chronic pain. These beliefs help to relieve some of the pain and boost their overall feelings of well being.

In an earlier study it was found that with hospitalized mentally ill patients, the mortality rate from cancer was much higher than it is for the general public; also the morbidity from cardiovascular diseases is higher. Prior to 1940 and the introduction of antibiotics, the morbidity from infectious diseases was much higher; the risk of contracting Parkinson's disease is higher; but the risk of being victimized by rheumatoid arthritis and allergic phenomena is lower.

What are we to make of all this? Medical science, faced with a very old, very common ailment, seems uncommonly baffled by it. Medical practitioners admit it is poorly understood. Many more women than men are afflicted and the medical profession seems not to have a clue as to why this is so.

One thing, however, they do know without question. Rheumatoid arthritis is an *inflammatory* disease, characterized by *inflammation* of a joint or joints. Unfortunately, patients suffering from it have not shared in a general improvement in life expectancy in four decades, according to a Mayo clinic study reported on February 26, 2000.

Osteoarthritis, the most common form of arthritis, alternately known as hypertrophic (abnormal enlargement of a part or organ), osteoarthrosis (bone-joint disorder), or DJD (degenerative joint disease), is a disease of aging, the "wear and tear arthritis."

Some sixteen million people seek treatment for OA every year, while many more suffer without seeking help. The main symptom is pain. Joint inflammation is rarely a problem until the condition becomes severe. OA usually begins after age forty and develops slowly, affecting only certain joints. By the age of seventy, it is present in almost everyone. It is not systemic and doesn't cause a general feeling of weakness.

OA is a breakdown of cartilage in the joints. First the smooth cartilage surface softens and becomes pitted and frayed. Over time it wears away completely causing the bones to rub together. The bone ends thicken and form bony growths, or spurs, causing pain and possible inflammation as the joint loses its normal shape.

The cause of OA is unknown, but several factors...metabolic, genetic, chemical, overuse of joints or injury...may play a role in its development. Obesity increases the risk of OA of the knee.

Treatment is based on how severe the condition is. The primary goal of medications prescribed is to relieve pain and to control inflammation if this is present. Medication may be taken on a daily or on an as-needed basis. These include non-narcotics such as acetaminophen, and narcotic analgesics such as codeine. Nonsteroidal anti-inflammatory drugs, or NSAID's, such as aspirin, ibuprofen or naproxen, may be taken to help reduce joint pain, stiffness and swelling.

Corticosteroids related to cortisone, a natural body hormone, may be injected into affected joints, but caution is required as they are known to have negative side effects. Topical analgesics in the form of creams, rubs or sprays may prove helpful in lessening pain.

Exercise is considered beneficial, also a healthy diet.

In the last few years, nutritional supplements have moved front and center. In their book *The Arthritis Cure*, the authors Jason Theodosakis, MD, M.S., M.P.H., Brenda Adderly, M.H A., and Barry Fox, Ph.D., claim that there is now available a substance that can halt, reverse, and possibly even cure OA. The nutritional supplements they are waxing so enthusiastic about are *glucosamine* and *chondroitin sulfates*, which are available in any health food store. These supplements are reported to have no significant side effects and used together work to regenerate cartilage and to keep harmful enzymes under control.

SAM-e (*s-adenosylmethionine*), a naturally occurring compound, is another nutritional supplement said to have been used in Europe for years to great effect in lessening the pain of arthritis, and MSM (*methylsulfonylmethan*), which occurs naturally in the body, is another supplement currently being put forth as helpful, or, in some cases, close to miraculous.

If all else fails, there is surgery. Damaged joints are repaired or replaced with wear-resistant artificial joints made of metal and plastic. Today there are replacements available for most of the major joints.

The third most common form of arthritis is gout. Whereas rheumatoid arthritis is predominantly a woman's disease, gout is a man's. It afflicts some two million Americans, 80% to 90% of them men over the age of thirty and closer to sixty years of age.

Gout is said to be a metabolic disorder. A poor diet can worsen the condition. Half of all cases begin in the first big toe joint. In gout, uric acid, a waste product in the urine cycle, is either over-produced, under-excreted, or both. When there is too much uric acid, some of it forms uric acid crystals. These crystals can end up in joint space, especially in the joint of the big toe, where they act as pieces of glass. The joint becomes hot, swollen, stiff and painful. The skin can appear red or purple and the pain is intense.

This particular affliction, among all the infirmities man's flesh is heir to, has often been seen as a joke...to those who don't have it. Is there any explanation for this?

In the chapter to follow, let's take a closer look at each of these three forms of arthritis...RA, OA and gout...to see if we can zero in on any overlooked causative factors. We will also attempt an answer as to why those stricken with gout...which is reputed to be a terribly painful condition...have historically been snickered at and made objects of fun.

15

Arthritis: Ready For Its Close Up?

❖

Dam Bones, Dam Bones, Dam Dry Bones

○ ○

Look now how mortals are blaming the gods…but in fact they themselves have woes beyond their share because of their own follies.

—*Homer*

The one thing positively known about rheumatoid arthritis is that it is an *inflammatory* disease, characterized by inflammation of a joint or joints.

Inflammation: The state of being inflamed.

COMMENTARY

Rheumatoid arthritis would seem to be an expression of the "state of being inflamed," of being angry or in a rage, a connection explored by researches in the mid-20th Century, but later apparently dropped.

Possibly we can see the connection simply by looking closely at the symptoms.

RA may come on suddenly or slowly. Anger growing into barely controllable rage may come on suddenly or grow more slowly.

RA may flare up, then fade away without a trace. Or it may keep returning until it is present virtually all of the time. Anger may flare up, be dealt with, and go away. Or it may flare up, be insufficiently dealt with and return. If it hardens into hate/resentment so that it becomes next to impossible to deal with, in time it may cling to its victim for longer and longer periods until it is present virtually all of the time.

If RA is a defense against the release of anger, a defense against a psychotic breakdown, then if a psychotic breakdown has already occurred, there is little need any longer to defend against it. Therefore, hospitalized, mentally ill patients would show a low incidence of it, as they do.

As the feminists have shown us all too clearly, boys and girls in our culture are subjected to entirely different social conditioning. Boys tend to be denied expression of many or all of the softer emotions. Softer emotions indicate to some parents a softer being, a non-butch, non-macho child, and who wants a sissy for a son? Anger is okay, a natural for macho kids, so with parents who react this way, boys are allowed their anger.

Girls, on the other hand, are often allowed all of the softer emotions but tend to be denied their anger. Young girls are expected to be little ladies, soft and sweet. Parents who think this way don't want a rough, rowdy, "unfeminine" daughter. As a result many women grow up unable to face their angers, inhibiting their resentments and their rage to a much greater extent than men.

Studies have shown that victims of RA are overly conscientious people who inhibit their rage. If RA is, as some research has indicated, an expression of repressed rage...if it strikes because someone has become so "burned up" that this inflammation is visited upon the joints...women would be much more likely to fall victim to it than men, as they do.

HYPOTHESIS:

Rheumatoid arthritis occurs when the mind becomes so full of repressed anger that it is no longer a proper fit for the spirit.
The spirit insists that anger is present and must be dealt with. The mind squirms away in fright refusing to accept the message, and the physical body is left to manage as best it can. An infectious agent...anger...triggers the disease. The body, absorbing the message that anger is there, mounts the only assault it can: Against itself.
The ill fit is expressed in precisely those parts of the body where the lack of fit is occurring.

The joints are constructed to move in a fluid, frictionless way. Resistance to movement causes friction, which produces heat, a proven fact of the physical world. Within the human body, which is part of the physical world, this fact holds true.

Resistance to movement produces heat, ie, inflammation. Isn't this the simplest explanation as to what causes rheumatoid arthritis, i.e., resistance to movement?

In RA the joints most often affected are in the fingers and toes.

We use our hands to reach out to others and often use them in work. If our feeling is, *I must do this but dammit I don't want to!* or *I want to do this but dammit I can't!*, we are resisting movement and causing friction/heat/inflammation.

We use our toes when we walk, i.e., go forward. If our feeling is, *I must go forward but dammit I don't want to!* or, *I want to go forward but dammit I don't dare*, we are resisting movement and causing friction/heat/inflammation.

Other joints frequently affected, and the way we can go about inflaming them, are:

Ankles: *I have to turn this way and that, even though dammit I don't want to.*

Knees: *I have to move forward but dammit I don't want to. Or the reverse.*

Hips: *I have to bow to others' will but dammit I don't want to.*

Wrists: *I have to turn my hands to this but dammit I don't want to. Or the reverse.*

Elbows: *I don't dare stiff arm/straight arm anyone as much as I might want to.*

Shoulders: *I have to shoulder this burden but dammit I don't want to.*

Neck: *I'm sick of having to bend to others, dammit, dammit, dammit.*

Jaw: *How I'd like to jawbone my way out of this, but I don't dare even try.*

An increasingly crippled body, to where the hands may become practically unusable and the feet badly distorted, to where tendons and muscles may no longer connect with bones so that movement becomes impossible, quite neatly solves all problems. What you can't do, you can't do, period...ending the war of resistance. But unfortunately, because this type of solution is not to the spirit's liking, the solution is experienced as an extremely painful one.

In addition to inflammation of the affected joints in RA, victims often suffer from the following:

Fever: A heating up of the blood, i.e., anger.

Weariness: It is tiring to inhibit rage.

Loss of appetite: Food=love. When we are caught in the grip of fury, we reject love.

Anemia: To be red-blooded is to be vigorous, fearless, daring. To rigidly hold in anger is the opposite of daring.

Lungs may be affected: How we draw in life/breath is always affected by strong emotion.

The skin, blood vessels, heart, eyes or muscles may be affected: Vital parts are always affected by repressed emotion.

Cold, tingling hands and feet: Fear, the flip side of anger.

Excessive sweating: Fear; the cold sweat of fear.

RA is clearly no fun, but possibly it is less painful than daring to stand up to others to release years of pent-up rage. When we are too frightened to go outward with our anger, where can we go but inward, turning on ourselves?

* * * * *

Let's look now at osteoarthritis, the most common form of arthritis, the wear-and-tear disease.

With aging, OA moves in to stay with almost everyone. Cartilage gives way, and joints become stiff and not as easy to move. With enough damage, bone can press against bone, causing great pain and on occasion inflammation.

Fortunately the wear-and-tear arthritis is a relatively benign disorder. Pain, when present, can usually be managed by medication and rarely becomes disabling. In severe cases, surgery is an option. Overall, OA can be an annoying problem rather than a severe disability for those healthy enough to journey into old age.

Although there are various theories as to what causes OA, it isn't known why some people get it at an earlier age than is usual or why some people make it through a long life with virtually no difficulty at all. Is there any reason for this variance in the age of onslaught and the degree of severity?

COMMENTARY

Emotional friction plays a part, the same kind of friction that in a more acute form results in rheumatoid arthritis.

If movement remains fluid and painless in the mind...in our thoughts and feelings...it will remain, for the most part at least, fluid and painless in the joints, regardless of any physical changes that aging brings.

HYPOTHESIS:

> *The pain of osteoarthritis comes not from rage against movement, but from the muted form of anger that is part boredom and part frustration. We have to continue to go forward, something in us says, while another voice in us objects, "What's the point?" We have to continue cooking, cleaning, walking, gardening...or do we have to? As we move closer and closer to the grave, shouldn't we be concerned with more important things, like, possibly, throwing off all shackles, all responsibility, before it's too late and really enjoying ourselves? Or climbing out of our rut and exploring ourselves and our world? Or at least doing something different and meaningful, even if it's only diving within to examine the state of our souls?*
>
> *In osteoarthritis, some joints can, on occasion, become inflamed. Here our anger is no longer muted, but is growing to where it catches fire, and inflammation results.*

People reach this stage of boredom and frustration at different ages, with varying degrees of emotional intensity.

* * * * *

The third most common form of arthritis is gout.

Arthritis: Ready For Its Close Up?   159

Hippocrates, the father of medicine, gave the first really clinical description of gout about 2,500 years ago. He outlined the patterns of the disease in three widely quoted aphorisms:

1. A eunuch does not take the gout.

2. A woman does not take the gout until her menses be stopped.

3. A young man does not take the gout until he indulges in coition.

Right there we have about all we need to know about the underlying cause of this condition. Also why this particular affliction has always seemed such a joke to those who aren't afflicted by it. In Western society, man, frightened by sex, has always tended to snicker and smirk about it.

According to medical science, the cause of gout lies in the way the body deals with excessive amounts of uric acid in the blood. Uric acid is produced by the body's normal metabolic processes and is discharged for the most part in the urine.

A man must choose between urination and sex. If he chooses sex too obsessively, and with too much guilt, hello gout.

The large joint of the big toe is commonly the site of the first attack of gout...for some unknown reason, according to the doctors. This should not be an "unknown reason" for the reason seems clear enough.

"Joint" is one of the slang terms for penis. Sex is a physical activity, largely performed by the physical part of the body (from the waist down), and the most physical part of this physical part of our bodies, the part that connects us directly with the physical earth, is the foot.

For this reason as well as others, the foot has always been closely associated with the erotic. It is equipped with extremely sensitive nerve endings, and the involvement of the foot, in particular the toes, and most particularly the big toes, in sexual practices is far from unknown.

Toe sucking is indulged in and advocated by the sexually sophisticated. "Don't knock it until you've tried it," remarked one young woman with earnest belligerence when the practice was under discussion. One particular royal princess of Britain got into serious trouble because she was caught by a photographer having her toes sucked by a man not her husband.

The big toe can be and occasionally is used as a substitute organ of copulation, primarily as a teaser.

If there is a build-up of sexual guilt to where the pressure to punish becomes intolerable but the guilty one (self-perceived guilty one) can't bear to risk his precious penis, what is the body to do?

The penis can justifiably claim that it provides a life and death service, urination. Also that it is not in itself responsible for sexual excess, which begins in the head, reducing the penis to the role of helpless slave. And because the penis does have a redeeming physical value and while sexual excess may deserve punishment, it nevertheless isn't a life and death matter, therefore it (the penis) should be excused from the block and a substitute put in its place.

What more natural substitute than the big toe?

The affected joint (the big toe in most cases) becomes unbearably tender, with swelling and redness. This is typical engorgement preparation for intercourse, but instead of being accompanied by pleasurable feelings, the pain is excruciating, standard aversion therapy, a fitting punishment for real or imagined sexual sins.

Gout is primarily a man's disease, but eunuchs don't get it, nor do young men until they have indulged in sexual intercourse. Women are protected until their menses have stopped and in a procreative sense (and in this sense only) they are no longer the women they once were. Gout is commonly thought to be, and studies have shown it to be, an affliction of high living, a disease far more frequent among the affluent than among the poor, and naturally it's all a big joke to those unaffected by it.

SUMMARY:

All forms of arthritis are caused by a build-up of repressed negatives...anger, boredom or guilt...to where the mind/body becomes an unfit habitation for the spirit.

HYPOTHESIS:

> *All disease...whether infectious or non-infectious...is caused by a lack of fit between mind and spirit.*
> *Every disease manifests in precisely that area of the body where the lack of a proper fit is occurring. This means that a knowledgeable reading of the body will tell us the why of any dis-ease.*

* * * * *

Before we turn to a discussion of how we can learn to read our bodies and lessen the chance of disease, let's have a look in the following chapters at an affliction that is currently causing tremors in us all: Cancer.

16

Cancer, The Modern Plague

❖

One?...One Hundred?...Two Hundred?...Three Hundred?... How Many Plagues Are We Up Against?

o o o o o o o o o o o o o o o o o o o o o o o o o o o o o o
Calamity is the perfect glass wherein we truly see and know ourselves.

—*William Davenant*

In 1971 President Nixon declared a war on cancer. If we threw massive amounts of money at the problem, we would have a cure within five years, he predicted. Thirty years later what has been accomplished? Over $35 billion has been spent on research, $1 trillion on therapy, and seven million people have died...with still no letup in sight.

For the year 1998...the most recent year for which figures are given on the Internet as I write this...it was estimated that 1.2 million Americans would be diagnosed as having cancer and 564,000 would die. If this gloomy prediction proved accurate, the medical profession is no longer achieving the cure rate claimed in 1931 of one in three. This one out of three cure rate was considered a failure back in the 1930s.

Notwithstanding these dismal statistics, when David S. Rosenthal, president of the American Cancer Society, was asked in February 1998 whether we were winning or losing the war against cancer, he responded that we were winning. If this is winning, what would losing be like?

The 1998 prediction was a worsening of both cases and deaths per year over the mid-1980s, when only a million people a year were diagnosed, with an expected death rate of about 440,000.

Cancer statistics indicate that currently about one out of every four deaths in this country is due to cancer. In December 1990 the United Press International news wire reported that cancer had overtaken heart disease as the leading cause of death in the United States between the ages of thirty-five and sixty-four.

Recently there has been a slight decrease in the cancer death rate. Most of this is attributed to changing nutritional habits.

We voluntarily donate money for cancer research and are taxed to support additional research, with a rising curve of new cases and a cure rate lower than that achieved back in the 1930s as our reward.

Why aren't we doing better in our war against cancer?

Cancer has become so prevalent that almost all of us have had some personal experience with it, if not in our own bodies then in the bodies of our loved ones or friends. Estimates are that one out of every three Americans will be diagnosed as having cancer at some time in their lives, and two out of every three families will have a member who has been stricken.

Why has our costly war against this plague proved so futile?

Decades of research, costing enormous amounts of money, haven't increased the cure rate. Today we are more vulnerable than ever. More of us are diagnosed with cancer and die of it every year. Therefore the expected mortality rate must be plotted on a rising curve. Even more of us will die of cancer next year than this, even more of us the year after that, unless we derail ourselves from the track we are on.

What if anything will save us?

"Cancer" is not just one disease, according to the medical profession. Rather it is a term used to denote a group of diseases characterized by the disorderly and uncontrolled multiplication of abnormal cells in the tissues of the body.

Primary types are:

Carcinoma, which arises in the epithelial tissues, that is, any tissue that covers a surface, lines a cavity or the like, and performs protective, secreting or other functions, as the epidermis.
Sarcoma, which arises in the connective tissue.
Leukemia, cancer of the blood.
Lymphoma, cancer of the lymph nodes.

While the cause or causes of cancer is or are not yet known, numerous substances are believed to be or have been shown to be carcinogenic, i.e., a substance that tends to produce a cancer. Primary among these substances is tobacco. Tobacco smoke is considered to be responsible for lung cancer and is believed to contribute to other cancers as well.

Other carcinogens are said to abound in our environment: Asbestos, vinyl chloride, tars, pitch, certain industrial oils, x-rays, radium, other radioactive materials, and certain food additives, to name the most widely recognized.

Because of all these identified carcinogens, there is a tendency at times to see cancer as environmentally induced. However, viruses are also suspected in some if not all cancers. Age and sex play a part. Men are more frequently victims of some cancers and women of others, and for both sexes the older we get the more prone to cancer we are. Nutritional deficiencies may cause or be an important contributing factor in cancer of the liver and esophagus while over-exposure to sun increases the risk of skin cancer. There may also be a built-in susceptibility to certain cancers due to heredity.

Cancer cells are very primitive, tending to be undifferentiated blobs of protoplasm. Most of the tissues in adult organs have a very specific look, a very specific structure. Liver cells are built like and look like liver cells and kidney cells like kidney cells. In contrast to this, a cancer cell in the liver will tend to look like a cancer cell in the kidney, for once cells become malignant they lose a lot of their secondary characteristics and their distinctive anatomy.

Currently there is a belief that cancer cells can arise in either of two ways. There can be a genetic break wherein something goes wrong in the DNA or the chromosomes and daughter cells no longer replicate mother cells identically. Or, under some as yet not fully determined stimulus, a mature cell can go berserk and become aberrant.

Normal cells divide until they come in contact with neighboring cells, then they stop. Cancer cells have lost this "contact inhibition." They aggressively invade any tissue surrounding them.

The cancer cell is wild and asocial. It proliferates without restraint and soon destroys everything in its path. Other cells may proliferate inappropriately and produce a tumor. However, if the cells of the tumor compress but do not invade neighboring tissue, they are classified as benign, non-cancerous. If invasion of neighboring tissues and adjacent organs occurs, the growth is labeled cancerous.

Another characteristic of cancer tissue is that ordinarily it does not stay in one place but reproduces elsewhere in the body. While some cancer cells remain where they are born, most travel. They infiltrate the blood vessels and circulate. This process is known as *metastasis* (the transference of disease-producing organisms or of malignant or cancerous cells to other parts of the body by way of the blood vessels, lymphatics, or membranous surfaces). This dissemination occurs very early, during the first few days of a tumor's development.

Thus the two fundamental characteristics of malignancy are invasion of neighboring tissue and metastasis.

That bacteria might be involved in the causation of cancer has been suggested for decades, only to be rejected for decades by the entrenched medical establishment. Some have seen a conspiracy at work in this, a deliberate suppression of evidence.

Early pioneers...Dr. Michael J. Scott of Butte, Montana, Dr. Thomas J. Glover of New York, and Dr. Glover's laboratory assistant, Tom Deaken...experimented in the early twenties with an anti-cancer serum reportedly with marked success. Newspapers of the time gleefully reported that a cure for the cancer plague had been found.

In 1940 Dr. Glover, in collaboration with a surgeon, Dr. John E. White of Malone, New York, published a report detailing their successful treatment of 328 cancer patients from the early 1920s to 1939. They published their report privately after the National Institutes of Health decided that publication by the Public Health Service would wait upon a repetition of their work.

In their book *The Cancer Conspiracy*, authors R.E. Netterberg, M.D., and R. T. Taylor wrote that it seemed obvious to them that the historical confirmation of the bacterial relationship to cancer that Dr. Michael J. Scott had announced in detail in 1924 had been deliberately suppressed.

Another researcher, Dr. W. M. Crofton, Lecturer in Special Pathology, University College, Dublin, Ireland, published a book in which he claimed that a microbe was always associated with cancer. He claimed to have made vaccines from the bacterial forms that he then used with surprising success. His work was fiercely attacked, even though others had confirmed his results, and he was all but completely silenced.

Experiments conducted even earlier by a cancer researcher, Peyton Rous, suggested a possible bacterial or viral causation. Working with chickens, Rous was able to transmit sarcoma by inoculating with dead cells from a tumor or with a tumor extract. Two other sarcomas were successfully transmitted using tumor juices. In each case the characteristics of the original growth appeared in the induced tumors. Dr. Rous and his collaborators became convinced that the filterable agent was an extremely small microbe.

That cancer might be caused by germs was an extremely unpopular notion with those in power even back in 1910, so Dr. Rous was derided and ridiculed and his experimental results pushed out of view and quickly forgotten.

Bacterial involvement in cancer has apparently been confirmed by more recent research. Dr. James Hiller, an expert in electronic microscopy, stated categorically in a paper he published that he detected a microorganism in every type of human tumor or animal tumor he studied.

Dr. Virginia Livingston-Wheeler, a researcher in the field of cancer microbiology, reported similar findings. In 1965 she and a co-worker, Dr. Eleanor Alexander-Jackson, reported in a paper that they had found the chemical *actinomycin* in the urine and in cultures from cancer patients. The most likely source of this toxic chemical compound were bacteria from the *Actinmycetales* group.

Drs. Wheeler and Alexander-Jackson named this bacterium the *Progenitor-cryptocides* (ancestral hidden killer). This bacterium could be classified as a virus as its life cycle includes a state so small that it cannot be detected with standard microscope techniques but must be viewed under an electron microscope. In Europe, other researchers working with cancer patients isolated and observed a microbe with identical characteristics

Some of those working in the field still follow Dr. Wheeler-Livingston's lead. Dr. Harry A. Knopper wrote as recently as 1998 that cancer is caused by an infectious agent, a shape-changing bacterial microbe named the *Progenitor-cryptocides* by the late Dr. Virginia Livingston-Wheeler. Ralph W. Moss, Ph.D., a science writer who has spent more than twenty years investigating and writing about cancer issues, also endorses her work.

Others report that her claims have been disproved. No one has been able to produce a cancer by injecting animals with the organism she said causes it. There have been numerous cases where cancer tissues did not contain the organism. Also cultures from Dr. Wheeler-Livingston's lab, when grown in other labs, turned out to be common forms of staphylococci that inhabit the skin.

Dr. Wheeler-Livingston reported a second discovery, which had to do with *chorionic gonadotropin* or *choriogonadotropin*...CG, or hCG. HCG is a substance synthesized and secreted by the *trophoblast*, the layer of extra-embryonic ectoderm that chiefly nourishes the embryo or develops into fetal membranes with nutritive functions.

When the sperm enters the egg within the female body and the cells begin to divide, the developing embryo, with its chromosomes from the father, is a foreign body and therefore subject to attack. One of the functions of hCG is to ward off this attack.

This is accomplished through a high negative charge. Agents of the immune system also carry a negative charge, but of lesser magnitude. Like charges repel each other, so the hCG effectively protects the "foreign intruder," the new life, from being assaulted and destroyed by the body's immune system.

In 1976 this substance (hCG) was also found in human spermatozoa. If the sperm weren't protected in this way, they'd be attacked and routed as foreign invaders long before any sperm, successfully beating out all others, could ever complete the hazardous journey to the egg.

Various investigators have also found the presence of a CG-like substance in the serum of a great number of cancer patients. This suggests that cancer cells, like the trophoblasts and sperm, are also producers of CG or a CG-like substance. Further study confirmed this. Cancer cells *do* indeed synthesize the CG-like compound.

In 1902 a man named John Beard, PhD., Professor of Embryology at Edinburgh University, at that time the preeminent medical school in the English-speaking world, published an article in *The Lancet*, a leading medical journal, on the trophoblastic origins of cancer. He followed this up with the publication of a book, *The Enzyme Treatment of Cancer and Its Scientific Basis,* in 1911.

Beard believed that he had figured out the cause and cure for cancer. He had begun his career working in the Adirondacks as a researcher, studying a certain kind of fish. This fish had alternating life cycles, one sexual, one asexual. Later Dr. Beard extended his studies to mammals, humans in particular. In time he concluded that cancer was asexual generation, or a wild and crazy trophoblast without an embryo.

Cancer equaled pregnancy in the wrong time and place.

Author Susan Sontag, in her book *Illness as Metaphor*, mentions with obvious disapproval an ancient metaphoric connection between pregnancy and cancer. Possibly she is referring to Dr. Beard's belief, based on years of research, or maybe such a belief predates Dr. Beard and is far more ancient. Unfortunately she doesn't elaborate.

In any case, Dr. Beard's one hundred-year-old thesis is currently beginning to stimulate a lot of research interest.

Dr. Herman Acevedo is a biochemist who has spent over twenty years studying a possible relationship between hCG, or human chorionic gonadotropin, and cancer. He states clearly that he does not belief hCG has anything to do with the etiology, or cause, of cancer. In his view cancer can be produced by a hundred different things, most of them in our environment. His interest is in the process whereby normal cells transform themselves into malignant ones.

Dr. Acevedo believes that when a mature cell becomes cancerous, it reverts to an embryonic state. As it divides uncontrollably, it churns out hCG. This keeps the immune system from attacking and destroying it

Dr. Acevedo has shown that cancer cells are identical in appearance to cells from a fourteen-week-old fetus.

Healthy, mature cells are free of hCG. Embryonic cells use hCG as a shield to keep themselves free of attack. Dr. Acevedo's research has led him to believe that *malignant tumors use hCG for an identical reason.*

HCG is known to be critical to sustaining pregnancy. It is conceded by the medical establishment that one of its jobs may be to ward off attack by the immune system. That cancer cells also produce this hormone has apparently made little impression upon most cancer experts.

Dr. Pentti Siiteri of the University of California, San Francisco, concedes that cancer cells may indeed produce hCG, but as far as he's concerned, no one has shown that this has any particular relevance to the disease. Dr. Drew Pardol of Johns Hopkins agrees, citing the fact that there is no direct evidence that hCG is important in malignancy.

Dr. Acevedo claims that it *is* important. Cells protected by hCG immobilize the immune system. The high negative charge on these cells repels any attack.

The fact that hCG seems to be produced by all cancers suggests that cancer is not a group of one hundred to three hundred related diseases, as is insisted upon by conventional medical theory, but is one disease with many related forms. This means that there is an excellent chance that an effective anti-cancer serum can be developed.

If cancer is one disease, and cancer cells ward off assault by the body's immune system with hCG or a hCG-like protein carrying a heavy negative charge, then a serum capable of chemically neutralizing the hCG-protein would enable the body's immune system to do its job, to get at the mutant cells to attack and destroy them.

Trials of such vaccines are already underway. In one such trial, cancer patients are being immunized against hCG with a vaccine developed by Dr. Vernon Stevens of Ohio State University as a contraceptive. Vaccines like this one can promptly end newly formed embryos by depriving them of hCG. The study is designed to find out whether such vaccines will have the same effect on malignant tumors.

If the vaccine works, it could be used against cancers of every type. The prediction is that such a vaccine would be cheap and simple to use. The incidence of cancer could possibly be reduced to where cancer becomes, over time, as obsolete as the deadly plagues of the Middle Ages.

Any possible vaccine is still at least three years away. In the meantime, what kind of treatments are out there for those of us already stricken? And are they any improvement over the treatments offered in the past?

These questions will be dealt with in the chapter to follow.

17

Treatment: Conventional and Otherwise

❖

If At First You Don't Succeed, Try... Try... Try... Try... Try... Try Again

○ ○

There is no armor against fate;
Death lays his icy hand on kings.

—*James Shirley*

As all of us surely know by now, the conventional medical treatments for cancer offered in the opening years of the 21st Century are surgery, radiation and chemotherapy, all aimed at cutting out or killing off the malignant cells without destroying too much surrounding tissue.

At the start of the 20th Century, the only treatment was surgery. Professor John Beard, after his intensive study of cancer and his brilliant intuitive leaps, suggested an enzyme treatment. He prepared an enzyme that he injected into a patient with extensive laryngeal cancer. The tumor dissolved within a few weeks.

Publication of the case in medical journals incited interest, and between 1902 and 1915 there were dozens of cancer patients given this treatment. After Beard's book, *The Enzyme Therapy of Cancer*, was published in 1911, there were controlled trials in which patients were given pancreatic enzymes under academic supervision. Although there were some regressions of cancer and even some cures, the overall results were mixed. The treatment was considered extremely controversial.

Madame Marie Curie, world-renowned Polish physicist and chemist working in Paris, Nobel prize winner in both physics and chemistry, suggested that radiation therapy was a simple, easy way to cure all cancers, and interest in Professor John Beard's work withered away. He died in obscurity in 1923.

Later in the 20th Century, when radiation proved not to be a cure-all after all, chemotherapy was developed, rounding out the trio of treatments still available today.

Not everyone is satisfied with these treatments. Radiation burns surrounding tissue. Chemotherapy poisons the patient. Surgery mutilates. At times the cancer can be stopped, but not before the patient is so weakened that she dies. When my sister was dying of cancer, I confronted the doctor, accusing him of lying to her. He had repeatedly assured her that the chemotherapy treatments were bringing her cancer under control. His response was that he hadn't lied. Her cancer *was* now under control. The problem was that the treatment had so destroyed her organs that she had only a short time still to live. The treatment was successful...according to the doctor...but the patient died.

A somewhat more controversial approach than the three medically sanctioned treatments mentioned above is immunotherapy in which bacterial agents are used to stimulate greater immunity in the host body.

For a time we heard a great deal about interferon, a protein substance produced by virus-invaded cells that prevented reproduction of the virus. Interferon was being publicized as a possibly powerful new weapon in the fight against cancer after early studies found that it tended to inhibit the growth of tumors and to stimulate the immune system to work against them.

Later, however, it was reported that a two-year study conducted by Dr. Shelby Berger of the National Cancer Institutes and three co-workers found that interferon, instead of inhibiting the spread of cancer, may actually help the cancer cells to spread into normal tissue.

Other non-conventional treatments include Krebiozen and Laetrile, both of which have been pronounced worthless by the American medical establishment but continue to have their passionate and dedicated adherents.

Dr. Steven Durovic, a former assistant professor at the University of Belgrade, who was living in South America at the time, experimented with a drug that he had discovered and that he believed would cure cancer.

He brought his drug to the United States and was put in touch with Dr. Andrew C. Ivy, vice president of the University of Illinois and head of the Medical School. Dr. Ivy was also Executive Director of the National Advisory Cancer Council and a director of the American Cancer Society.

Dr. Ivy became interested in Dr. Durovic's substance, Krebiozen, and decided to test it scientifically. According to some accounts, the results were positive. The drug, though used only on cancer patients diagnosed as hopeless and close to death, helped remarkably. In some cases, pain disappeared. In other cases, tumors disappeared and were replaced by healthy tissues. Physicians experimenting with the drug across the country were having similar results, or so it is claimed. So what was the fate of this promising new cancer treatment?

Dr. Ivy lost his position at the University of Illinois and was asked to resign from both national cancer societies. His work was condemned as faulty and unscientific.

"A Status Report on Krebiozen" was published in the *Journal of American Medicine,* a report shot through with fraud, according to an investigator, Herbert Bailey, who wrote two books on the subject: *K-Krebiozen...Key to Cancer?* and *A Matter of Life or Death, The Incredible Story of Krebiozen.* The doctor who wrote the damaging article, according to Mr. Bailey, falsified the actual findings. Nevertheless this status report...valid or fraudulent, scientific or not...carried the day and, medically speaking, Krebiozen was dead.

Laetrile was another treatment which tried and failed. It was developed by a father and son team. Ernst Krebs Sr. was a physician. Both he and his son, Ernst Krebs, Jr., were aware of the pioneer work of John Beard and were convinced of the rightness of his view. They considered Beard's trophoblastic theory of cancer more than a theory. They considered it fact.

Beard among other things had been a brilliant embryologist. Some of his discoveries are still quoted in the embryological literature. He had a special interest in the placenta, which is produced after the fertilization of the egg.

To begin with, the newly fertilized human egg lives off its own substance, but there is not a lot of substance there so it must soon find a source of additional nutrition. As it passes from the fallopian tube into the uterus, it faces a second problem. To get the additional nutrition it needs, it has to somehow anchor itself to the uterus.

It starts producing a layer of tissue called the *trophoblast,* which very aggressively invades the uterus. John Beard had found that in every mammalian species he studied that the trophoblast invaded the uterus, developed into the placenta, and then in time stopped growing.

In those very few cases where it didn't stop growing, it led to a cancer called *choriocarcinoma.* For a long time this was one of the most deadly of cancers. Mothers would die of it within a few months. This cancer is one of the few that can now be easily controlled with chemotherapy.

In every species he studied, Beard found a definite time when the placenta stops growing. He determined that on the 56th day after conception, the placenta in the human mother stops growing.

He then began to study the question as to what caused the placenta to stop growing. He studied the developmental stages of the different organ systems of the fetus to see what he could find. In time he discovered that in every species he studied, the very day the embryonic pancreas becomes active, the placenta stops growing. From this he concluded that the pancreas was the single factor that controlled placental growth.

From this came Dr. Beard's belief that pancreatic enzymes were the key to the fight against cancer.

Dr. Krebs, Sr., and his son were convinced that Dr. Beard was right and that the way to cure cancer was through the use of pancreatic enzymes. Feeling that this alone was insufficient, they added something that they saw as an anticancer factor, a concoction made of apricot seeds known as laetrile, or vitamin B17.

A few doctors offered laetrile to their patients here in the United States, but the majority of those interested in the treatment traveled to Mexico in the 1970s to receive it.

Treatment with laetrile combined with a stringent diet was practiced in the clinic and hospital of Dr. Ernest Contreras in Tijuana, Mexico. Anyone journeying there, as I did in the mid-1970's, could easily find people returning for their semi-annual or annual checkups and drug supplies. These patients claimed they had been helped or cured. After receiving death sentences from American doctors, who pronounced them incurable, they headed to Mexico seeking hope and help. According to their enthusiastic testimony, they received both.

There were also patients who, though hospitalized and promised a cure, failed to respond and died. Their corpses were quickly and quietly whisked away. Figures as to how many were helped and how many weren't were not easy to come by.

Experts here at home were not impressed by those who reported being helped. Instead they warned that laetrile could cause cyanide poisoning. The Food and Drug Administration declared it illegal, which ended its peak usage by 1980.

Currently, according to a CNN report, laetrile is on the rebound on the Internet. Three firms are offering it for sale and are finding takers. A U.S. district judge in Miami, Florida, issued a preliminary injunction halting these sales.

Another approach that was controversial to begin with but has since become widely if not universally accepted is the Simonton method, developed by Dr. O. Carl Simonton of Fort Worth, Texas, who currently practices psycho-social medicine at the Simonton Cancer Center in Pacific Palisades, California. Dr. Simonton developed the first systematic emotional intervention for use by cancer patients. A pilot study conducted from 1974 to 1981 proved that it *was* of help, both in increasing survival time and in improving lives.

Dr. Simonton developed his method as he had come to believe that psychological forces played an important, possibly crucial role in the development of cancer and that these same forces could be rallied and used to overcome it.

To mobilize these psychological healing forces, he developed a special mental technique called *imaging*. The cancer victim was taught *healing visualization*, wherein he visualized an inner battlefield upon which healthy cells drove out or destroyed malignant ones.

In Dr. Simonton's view, there was a cancer-prone personality with identifiable characteristics. These characteristics included the inclination to cling to resentments, to be unforgiving, to indulge in self pity, to find it difficult to develop close, enduring relationships, and to suffer from low self esteem.

For Dr. Simonton, when he developed his initial program, no one "gets cancer." There is a wish to withdraw from life and in pursuit of this the choice is made to develop cancer. He saw cancer as a disease of despair, and malignancy as despair manifesting in cells

Predictably, when Dr. Simonton first developed his treatment approach, the response of conventional medicine ranged from labeling the method unproven to condemning it as unethical and dangerous. It was all too clearly a *"Blame the patient"* approach. Dr. Simonton admitted that his treatment approach could engender guilt, possibly hurting people more than it helped them.

In an attempt to prove or disprove an emotional aspect to cancer, a controlled study of 160 women admitted to King's College Hospital, London, was carried out by Steven Greer and Tina Morris. This study failed to confirm earlier reports of association between cancer and stress, in particular the loss of a loved one, habitual denial of stress, and depression.

The study also failed to substantiate Dr. Simonton's earlier contention that cancer is a physical manifestation of a state of despair. Also, since cancer is known to develop in many different species, dogs, reptiles and fish among others, and even to plants, it seems implausible to many that emotional factors would predominate over biological ones.

Nevertheless, addressing the mind, emotions and spirit of cancer patients is now viewed by most practitioners as an important part of overall treatment. Relaxation classes using imagery, visualization and light hypnosis are offered in numerous cancer centers.

At any given time, dozens of other therapies are offered, most of them out-and-out quackery.

As to preventive measures to avoid cancer, one clear warning we are given on all sides is against the use of tobacco. Nutritionists also warn against the additives in prepared foods, the hazards of our impure drinking water, and hypothesize that cancer may be a deficiency disease, or a disease brought on by a lowered immune system. They claim there is evidence that some important nutritional substance is lacking as cancer progresses.

On Sunday, December 13, 1981, the Baltimore Sun ran a story by Isaac Rehert reporting on the experience of one man who attempted a nutritional cure for his cancer

Dr. Anthony Sattilaro, a fifty-year-old chief executive of the Methodist Hospital in Philadelphia, learned in his annual physical checkup in 1978 that his body was riddled with cancer. Three operations were performed in three weeks. He was put on large dosages of estrogen and painkillers were prescribed to deal with his near constant pain.

His father was also fighting cancer, and in early August that year he lost the battle and died. On the trip back from the funeral, feeling depressed and weary from pain, Dr. Sattilaro picked up two youthful hitchhikers. When he mentioned that he was dying of cancer, one of the young hitchhikers, Sean McClean, responded that he didn't have to die. Cancer wasn't that hard to cure. All he had to do was go on a macrobiotic diet, a diet consisting of whole grains, vegetables, beans, fish, soup, fruits, seeds and nuts.

Figuring he had nothing to lose, Dr. Sattilaro decided to give it a try. After going on the diet, he began to feel better. In time, against medical advice, he stopped taking the estrogens, and continued to feel fine. In September of 1979 he was checked again and there was no sign of cancer anywhere in his body.

A nutritional approach that worked...for one man at least.

Are there other approaches that might help, or theories about the cause of malignances that could guide us?

To try to answer this, in the following chapter let's take a closer look at two specific cancers, breast and lung.

18

Cancer: The Mutant Cell

✦

Under Too Rigid a Tyranny, Do Our Cells Rebel and Start the Revolution Without Us?

○ ○
Extreme fear can neither fight nor fly.

——*The Rape of Lucrece*
  Edward de Vere, 17th Earl of Oxford"
  (Shakespeare)

We have already hypothesized that all diseases...infectious or non-infectious...have only one cause: A lack of fit between mind and spirit. Whether cancer is a non-infectious disease, as is currently believed by most oncologists, or will in time prove to be a low-grade infectious disease, as some researchers now claim, is immaterial to our argument. When a misfit occurs, the spirit can surely summon whatever means work best to bring the lack of fit to the mind's attention. We don't have to know the particular means used in any given disease.

Researchers have found that a protein substance CG (chorionic gonadotropin or choriogonadotropin) or a CG-like protein substance is found on the surface of spermatozoa, on fetal cells, and on cancer cells.

The sperm are protected, and the embryo is protected, to ensure the safety of new life. For some reason, malformed, mutant, uncivilized cancer cells are put under the same umbrella of chemical/electrical protection provided for the sperm and the fetus. Cancer, no matter how destructive it may prove to be, is treated within our bodies as a welcome invasion of new life.

Or, as some researchers put it, cancer can be looked upon as being an unwanted pregnancy.

HYPOTHESIS:

> *From birth on, we produce cancer cells, but cancer as a disease develops only when the tyranny under which the cells live has grown so unbearable that rebellion comes as a welcome relief, a welcome manifestation of new life in protest against the old.*

* * * * *

To see how this may work, let's take a close look at one of the leading causes of cancer deaths among women, breast cancer.

Breast cancer is the second most common cancer among women, skin cancer being the most common. It is also the second leading cause of cancer deaths in women, after lung cancer. Medical authorities don't know what causes this cancer, but certain risk factors have been identified. These are:

Gender: Being a woman is the main risk.
Age: Aging makes us more vulnerable.
Genetic risk factors: About 10% of breast cancers are linked to mutations in certain inherited genes.
Family history: Having a close relative...mother, sister or daughter...who had or has breast cancer almost doubles a woman's risk.
Personal history: Any woman who has or has had cancer in one breast is at risk for developing a new cancer in the other.
Race: White women are at a higher risk than African-American women while Asian and Hispanic women run the lowest risk.

Breast biopsies: Abnormal biopsy results can be linked to a slightly higher risk.

Radiation Treatment: Chest area radiation treatment of a child or young woman can significantly increase her risk of breast cancer later.

Menstrual periods: Early menstrual periods (before age 12) and relatively late menopause (after age 50) pose a small risk. The same is true for women who have never had children, or had their first child after the age of thirty.

Other factors can increase the risk of breast cancer slightly: Using birth control pills, hormone replacement therapy, failing to breast feed, alcohol use, being overweight, failing to exercise.

There is no evidence to suggest that antiperspirants or bras can cause breast cancer, despite one book that claimed to find a causal link between the wearing of tight brassieres and cancer of the breast.

Smoking, induced abortions, breast implants and the effect of the environment are all factors that have an unknown impact on the risk of this cancer.

It is difficult to sort through all these risk factors and come up with anything definitive. This may *increase* slightly, this may *decrease* slightly, the risk we run as women of getting this disease.

How can we best zero in on what causes this dread disease and possibly figure out...apart from being born male...the surest way to avoid getting it?

First, we can go back to basics and remind ourselves of what is absolutely known about cancer in any form, which is:

Cancer cells are mutant cells, rebellious cells. Mature cells go berserk and regress to immaturity, or daughter cells go wild and stop replicating mother cells. New life stirs and grows aggressively...a pregnancy at the wrong time in the wrong place but new life nonetheless.

What causes rebellion?

Historically rebellions have been staged when a government becomes unbearably repressive, when the governed either have their potential thwarted and suffer from non-use, or are wrongly used and suffer from misuse.

To apply our earlier hypothesis specifically to breast cancer:

HYPOTHESIS:

Cancer of the breast is a rebellion against non-use or misuse of the breast.

COMMENTARY

The female breast serves two closely related functions. It is a sexually attractive feature. It is the organ that produces and secretes milk for the nourishment of the young.

Biologically speaking, the former function is subservient to the latter. That is, biologically speaking, the purpose of being sexually attractive is to attract a mate in order to propagate.

To phrase it another way: The breast has only one primary function and that is to produce and secrete milk for the nourishment of the young.

The female breast, uniquely among all bodily organs, is a source of food, of nourishment for the new life the body has brought forth. Most of us associate food with love, which means the breast becomes a symbol of maternal love. If the breast is not allowed to fulfill this function, if maternal love is rigidly repressed, a rebellion of mutant cells may be staged as a measure of last resort, the only hope of throwing off the tyranny of repressive fear.

Women who have never had children run a slightly increased risk of developing breast cancer. Women who delay childbearing until after the age of thirty run a slightly increased risk. For a variety of reasons, women who have borne children may find it too painful to love them freely, or may be so engrossed in other endeavors they repress their longing to love them. All of the above cases can be considered the non-use of the breast.

Misuse of the breast can occur as follows:

Though there has been a great deal of ferment lately and some change, most women in our society are still being culturally conditioned to play a passive, subservient, sex-object role vis-a-vis men. This is in sharp contrast to the maternal role, which ideally is an active, protective role, a role similar to that which the male in our society is conditioned to play vis-a-vis the female.

In many ways the maternal role calls for traits and behavior which our society, in all its wisdom, has labeled "masculine."

If a woman, regardless of how many children she has given birth to, is afraid of this active, protective, "masculine" role, if she uses her breasts as a sexually attractive inducement but in fear never moves from this passive, subservient, sex-object role into a more active, dynamic, "masculine" role in order to freely love and protect her children, she is in effect allowing the secondary role of the breast to predominate over the primary role. This constitutes misuse.

Fear of this more active, dynamic, "masculine" role...a fearful clinging to a more passive role, leading to an over-dependence on men and on male approval...is usually caused by, or translates into, the fear of being considered lesbian, ie, the fear of lesbian feeling.

With a minimum of introspective effort, most women know whether they feel basically independent of men or overly dependent upon them, though this may not be perceived as a yardstick upon which to measure fear of latent lesbian feelings.

Years ago on a TV game show, "Tattletales," three married women were asked if there was anything in life more important to them than their husbands' approval. Two said no, there was not. The third said her husband's love was more important to her than his approval. Not one of them said that *self-approval, self-love,* was more important.

Women have been socialized to express feelings that they often do not really have, and the three wives referred to above may simply have been responding by rote to what has been culturally fed into them. In addition, this telecast was possibly two decades ago and women are slowly but surely changing. However, there are still a great number of women who cannot feel good about themselves, who feel rotten and depressed, isolated and out in the cold, lost in the darkness, if they are not basking daily in the sunshine of male approval. This degree of dependency is an affront to the dignity and power of the female breast and constitutes abuse. Any woman that dependent on men should look within herself to find out why, to face the underlying angers and fears that keep her in such an enslaved state.

Any woman who trembles at the thought of losing her lover, who will compromise herself or the welfare of her children to cling to some man, should reconsider her priorities. Any woman who makes the highly charged claim that she would rather see her beloved dead than in the arms of another woman may be sitting on a powder keg of repressed fears. Women who don't feel right about themselves without some man at their side should attempt to deal with these fears before they harden into rebellion in one or both of her breasts.

To reword and make even more specific our hypothesis:

HYPOTHESIS:

Breast cancer is caused by a severe repression of maternal feelings, a craven over-dependence on male approval, or a severe repression of lesbian feelings.

The spirit sends a message demanding change, demanding that fear be faced and dealt with, that all feeling be allowed to flow freely. This message is sent to the mind, but the mind darts away, refusing to listen, and in time the body is sufficiently impressed to respond, "You want change? We'll give you change!" The cells rebel, grow wildly and a cancerous growth results.

Cancer of the breast proclaims, in effect, "You failed to use your breasts properly. Therefore, we, your rebellious daughter cells, will now take over to run things ourselves."

These wild, rebellious cells have a profound, unassailable logic on their side. If we fail to use or misuse anything, how long can we legitimately expect to remain in control? Cancer signals a loss of control as a new life takes form.

One psychoanalytic researcher, Claus Bahnson, noted that women who develop breast cancer tend to be conformist, with strong religious beliefs.

In our culture women are still expected to be somewhat passive, receptive, to be sexual objects, non-aggressive. This cultural pressure finds its roots in and is reinforced by many religions. The more a woman conforms to sexual stereotypes, and the more seriously she takes her religion, the more difficult it will be for her to fulfill the maternal/masculine function of her breasts, and the more susceptible she will be to a cancerous rebellion.

* * * * *

Whereas breast cancer was for decades the leading cause of cancer deaths in women, currently the leading cause of cancer deaths in both men and women is lung cancer.

Cigarettes have been indicted as a leading causative factor in lung cancer. In fact, it is common to run across statements in print that cigarette smoking "causes" lung cancer, which is a very broad if not careless use of the word "causes."

While a link between smoking and lung cancer has long been established, this link is very clearly no direct cause and effect relationship. We all know people who have smoked for fifty or more years and who die peacefully in their beds at ninety, and others who have never smoked yet who die of lung cancer at sixty-two.

Phenomena can occur concurrently without there being a causal relationship. When we have a cold, we are often bothered by sore, itchy eyes as well as a runny nose. The soreness in our eyes does not cause the nasal discharge or vice versa. Both are symptoms of the same underlying cause, the cold we have.

Cigarette smoking and lung cancer, like itchy eyes and runny noses, may be linked only in that both are symptoms of a deeper underlying problem.

COMMENTARY

Primitive peoples often eat organs of choice from the bodies of those they have slain as a way to incorporate into themselves the admirable qualities of their fallen enemies. In the Catholic Mass, the flesh of Christ is ingested in symbolic form, and his blood imbibed.

Cigarettes are phallic in shape. When we smoke a cigarette, we are symbolically taking into ourselves the phallus.

But this isn't all we are doing. As we smoke and take in the phallus, we are at the same time destroying it through fire. As we smoke, we are expressing an unresolved love-hate relationship.

If there is a deep dissatisfaction over gender identity, if fury or fear is at work, we can express it by becoming heavy cigarette smokers, thus risking cancer, or by developing a growth or growths in the lungs without ever smoking. Neither one causes the other. Both are symptoms of the same underlying dissatisfaction, fury or fear.

Gender identity is a fundamental identity. The moment a baby is born, even before it is slapped or otherwise prodded to take its first breath, the sex is noted. The first thing we ask about a new arrival, once we hear that both mother and child are alive and well, is: *Boy or girl?* Therefore problems with gender identity, a fundamental identity, express themselves in the organ that processes the most fundamental requirement in life, breath, the breath that is life.

HYPOTHESIS:

Lung cancer is an expression of anger/fear over gender identity problems.

To phrase it another way: Lung cancer is the body's response to a craving for increased masculinity, or a less vulnerable masculinity, a craving caused by anger/fear.

The spirit shouts, "Relax! Be yourself! Stop being so afraid!" The mind darts away, refusing to listen, but in time the body is sufficiently impressed to respond. Lung cells rebel and go wild, resulting in a cancerous growth.

* * * * *

The dissatisfaction and anger that some women feel at not being men, at not having the rights and prerogatives that still adhere to the male in our society, can be fierce enough to make heavy, chain smokers out of them. But if this heavy smoking sufficiently expresses the fury...the two dozen times a day the woman ingests the penis, taking in penile attributes, while at the same time destroying an endless chain of penises, burning them up, blowing them out contemptuously in smoke...then the smoker may escape any malignant growth in the lungs.

If the smoking does not successfully express the full scope of the anger, with its reverse face of fear, then the remaining fear and anger must be dealt with, quite possibly through a cancerous growth...an unwanted pregnancy...in the lungs.

Recent reports indicate that women are now falling victim to this cancer at an alarmingly high rate. For years lung cancer was considered a man's disease, but no more. Women now account for 39% of lung cancer deaths, double the rate of 1965. Another disturbing fact is that while the survival rate for many once fatal cancers has improved dramatically, the lung cancer survival rate has hardly budged, moving from 13.4% to 13.9%, according to a Los Angeles Times report dated March 26, 2001.

Smoking expresses the same anger/fear for the male. But even as the heavy male smoker ingests a great number of penises to boost his uncertain gender identity, at the same time he is destroying, burning up, the very increased penis-power he feels most in need of, which means that smoking can exacerbate the underlying fear quite as much as it eases it.

For both sexes, smoking is an angry protest against gender expectations. The male smoker is saying, in effect: *If only all this hogwash about the "masculine" would go up in smoke, then I could stop my incessant worry that I'm not manly enough, not "masculine" enough.*

Both sexes are protesting the myth of masculine "superman" invincibility. Women in general see through this myth and are infuriated that it lingers on, giving men the edge in traditional masculine pursuits. For men, the protest is both intense and personal. While both sexes are to some extent victimized by the myth, the male is subjected to somewhat greater stress.

That our cultural stereotypes of the superior male place men under greater stress is statistically borne out in several ways. Men die at younger ages, more often falling victim to stress-related illnesses. More men commit suicide, become criminals or alcoholics. It may not be easy to be a woman in our society, but evidence from every direction proves it is even more of a strain to be a man.

To research the question as to whether stress and cancer are linked, Robert J. Keehn of the Medical Follow-Up Agency of the National Academy of Sciences National Research Council studied the health records of former World War II and Korean war prisoners. He failed to uncover any heightened cancer death rate.

If stress and cancer are linked, if stress contributes to cancer, then wouldn't these former prisoners of war, subjected as they were to long periods of abnormal stress, have fallen victim to the disease at a higher rate than the general population? The study showed they didn't so any stress-cancer link was considered disproved.

The above study no doubt was a worthwhile one, but there are surely varying kinds of stress.

One stress that would *not* be increased under prisoner of war conditions is the stress caused by uncertainty about one's proper gender behavior.

For all the deprivation and fear suffered by prisoners of war, one uncertainty they are not up against is what constitutes, in a prison camp, proper "masculine" behavior. Instead there would be reasonably clear, straightforward standards as to what it means, under those conditions, to "be a man" or to "act like a man."

But what does it mean to "be a man" or to "act like a man" in our "free" society? Even with the advancing tide of women's liberation, men are still expected to work and provide, or help provide, support for their wives and children, to be "good providers." So how much should they lie, cheat, and trample on the rights of others to achieve success as providers? How much should they "go along" to "get along"?

In his delightfully entertaining book, *Donahue, My Own Story,* Phil Donahue wrote that the biggest shock to him of growing up male in America was learning how often a bread-winning male was expected to kiss ass.

If it's necessary to kiss ass to get along, how much ass kissing is enough? How much is too much? At what point should a man stop sacrificing his self-respect as he struggles to live up to his expected role of good provider? *Help, someone, I don't know the rules. Where do I go to learn them?*

Even in our day, at the start of the 21st Century, after decades of effort on the part of some to achieve "liberation" for everyone, men are still expected to wear the pants in the family, or at least one pair of them even if his wife wears another pair. Few types in our society are as laughed at, sneered at, looked down upon as the milquetoast male, the man who is perceived as being at the mercy of a balls-busting wife, girl friend or mother.

Seared into the gut of every boy growing up in our country is this contempt for the "weak" male. Manliness is all. The favorite term of derision that young boys throw at each other is "*faggot!*"

Not long ago the papers reported another school shooting. A small, slender, fifteen-year-old boy, a good student who was apparently teased by his schoolmates and made the butt of their jokes, responded by grabbing a gun and splattering bullets around, killing two and wounding thirteen. In striking back, turning the tables on those who were persecuting him, did he "prove" his masculinity, become a "man," or did he merely "prove" how weak he was, how unable to live up to the masculine myth?

In the rash of school shootings we've been plagued with recently, how often had the gunmen been forced to absorb one too many taunts about their masculinity, their lack of manliness?

Most boys, even those who fail to fit the teenage image of macho manhood, manage to grow up without any violent acting-out, but at what cost in self-doubt and suppressed rage?

After a boy seething with inner anger grows up, marries, and possibly has children, if his wife argues with him, questions his judgment, at times even overrides him, how soon will anger…fear…worry…set in? Is he becoming a joke? Is the most important person in his life disrespecting him? If she is, what should he do? Throw up his hands in despair and walk out on her and the kids? Demand a divorce? Find himself a girl friend and have an affair to prove his manliness? Or maybe bust her across the mouth a couple of times to show her who's really boss? Or will that merely prove he's a bully, a no-good scum? *Help, someone, I'm not sure of the rules. Where do I go to learn them?*

Maybe his home life is fine, but there's a problem at work. Say he's just gotten promoted with a substantial increase in pay. His wife is delighted, the kids excited, and he feels wonderful. Now they'll be able to pay for Jimmy's braces and for Lori's piano lessons. But the new job brings with it a new boss or supervisor, and within a day he's fighting despair. His new boss constantly puts him down, humiliates him in front of his co-workers, makes him the butt of endless jokes. What should he do?

Should he try to ignore what's going on and persuade himself it will go away? Say the hell with it and walk off the job? Confront the boss and tell him to knock it off, risking unemployment and a rap as a troublemaker? Or ask to be demoted back to his old job where his co-workers liked and respected him and he enjoyed the work?

This is what he'd like to do but he'd have to explain it to his wife, and he shudders at the mere idea. He'd feel like a wimp and a crybaby. Men are supposed to be able to take it, aren't they? *Help, someone, I'm not sure of the rules. Where do I go to learn them?*

There are thousands of scenarios like this, wherein men find themselves feeling angry, and under stress. How easy to reach for a cigarette, light up, inhale deeply, and feel some of the churning rage let up a bit.

Surely few things can be more stressful than to be forced to play an extremely serious game, a game in which one's self-image and reputation will be forged, without being given an adequate understanding of the rules of the game. During the last century the relationship between the sexes was in such constant ferment that no one could possibly figure out the rules. All of us, men in particular, continue to live under a heavy barrage of ambivalent messages, to the point that it is no wonder a certain percentage of us loathe the very idea of "masculinity," the "manly" man.

If we live with enough churning fury over it, we can make ourselves sick. For men, there is surely a hope that if only they could get enough of the stuff...prove to be enough of a "man"...they could then turn the tables on the damn masculinity concept and destroy it, as they would happily give their right arms to do, so they grab for one more penile-shaped cigarette, and then one more, and then one more.

Smoking expresses this anger/fear, but unfortunately it often costs a man his lungs and his life instead of an arm.

Not all symptoms of any given affliction are always present. It is possible to get a severe cold without suffering from a sore throat or developing a cough. People can struggle with a great deal of unresolved anger/fear about gender identity and gender expectations and yet for one reason or another never take up smoking. The only symptom expressing their underlying rage may be a malignant growth in the lungs.

With our passage into the 21st Century, surely we can feel genuine hope that the incidence of lung cancer rates will drop. More and more the world is loosening up and stereotypes are fading away. The old days when men were expected to be MEN and women were assumed to like it are being seriously challenged and in some quarters are going down to defeat. Both sexes seem slowly more willing to allow men to be human as opposed to being hard, erect, silent, stoic, money-making penises, while women are more and more winning their struggle to be accepted as fully human too.

When women feel a deep down satisfaction with being women, no longer are frightened or angered by it, accept it fully, they will either not smoke at all or will quit the habit. When men find it possible to relax into their own masculine gender without fury or fear, fully accepting themselves, they will either not smoke at all or will quit the habit. And when these conditions are met, when both men and woman can relax and enjoy, lung cancer will be no more.

Let's turn to a discussion now as to the best way to achieve or maintain physical health in this frantic, fearful world.

19

Body-Squeak...Body-Speak

❖

If Our Bodies Ran on Wheels, Would We Heed the Squeaking-Squealing?

o o o o o o o o o o o o o o o o o o o o o o o o o o o o o o
God changes not what is in a people, until they change what is in themselves.

——*13:11*
The Koran

Most Americans take better care of their automobiles than they do of their bodies. We chow down fast fatty foods so relentlessly that 60% of us are overweight, and the percentage of us who are is increasing. We not only over-eat. We work too hard, play too hard, and worry too much. We stay in a near constant state of anxiety trying to own things that once owned won't satisfy us, and chasing after a happiness or peace of mind that continually eludes us.

For the most part we pay no attention to what our bodies are saying to us but instead drown out any and all warning signals by downing pills.

In contrast, a car is ordinarily taken in for regular tune-ups/check-ups, if the owner can possibly afford it. Pings and squeals are listened to, and every attempt is made to keep the machine running at maximum efficiency. Few car owners, in response to a warning signal, would rely on lifting the hood and tossing a couple of pills down the radiator, secure in the feeling that within a few minutes relief will arrive and all problems will magically go away.

This mystical cure-rite is reserved for the physical body, one's car being far too important to be treated in this slaphappy fashion.

There is today such a wide array of medications to alleviate human ills as to stagger the imagination. No matter what area of the body hurts, there is a pill to assure relief.

If you have a headache, pop a pill.

A stomach ache? Heart-burn? Pop a pill.

A backache? Here, down two of these.

Eyestrain, earaches, discomfort in your nose, throat or sinuses, cold sores on or in your mouth, a cough, pain in any part of your torso or limbs, don't worry about it, don't feel concerned. Your kindly, prescription-writing doctor or friendly neighborhood druggist can give you something to keep you from hearing your body's squeaking/squealing.

If you're feeling anxious, frightened, lonely or depressed, cheer up. There are pills for that too, tranquilizers or anti-depressants to soothe and comfort you, to make the squeal go away. In this day and age, no one has to listen to the warning signals his body sends out. We can muffle them, stifle them, chemically knock them out, somehow persuading ourselves that this will do it. Our pills will not only bury the symptom but will somehow, through magic, kill off the underlying cause of the symptom as well.

Inscribed in letters of gold on the temple at Delphi, the site of the most famous oracle of the ancient world, was the admonition *Know Thyself.*

"Knowledge is power," said Lord Bacon, and certainly the more thoroughly we understand and know anything, the greater is our power and our control.

COMMENTARY

If we would keep our thoughts and our feelings in proper fit with our spirit, we need to dive deep within to get to know what is boiling away down there. *Know Thyself.* While this is easy to say, it is not the easiest job in the world. It may be the hardest.

The Bible tells us that perfect love casts out fear. The reverse is also true. Where one exists, the other is banished. Love casts out fear. Fear casts out love.

It seems safe to assume that almost all of us live somewhere along the scale between perfect love and complete and abject fear. If we lived totally at the love end, one hundred percent loving without fear, we would be living saints, too light, too good for this world. If we lived totally at the fear end, one hundred percent afraid without a glimmer of light or love, we wouldn't last long enough to worry about it. The weight of the fear would crush us. So somewhere between the two extremes of total love and total fear, we struggle along, striving and crying, expressing, repressing, and generally making a botch of things.

HYPOTHESIS:

> *Love is a sense of unity/oneness, while a sense of separation...isolation...alienation...equals non-love, or anger/fear.*
> *A movement toward love fits the spirit. A movement away from love, toward increased anger/fear, ill befits the spirit and causes strain, a strain that can in time result in physical disease.*

Or to put it another way:

> *All disease is caused by the unwillingness...or, more precisely, the inability...to love.*

To put it even more simply:

> *Love is the basis of good physical, mental, emotional health.*

Or,

Love is good health. Fear is the absence of good health.

Pain is the language the indwelling spirit uses when speaking to us, to get the message across to us.

* * * * *

If we'd like to stay healthy by becoming less fearful and more loving, what is the best way to go about it?

To become a thoroughly good man is the best prescription for keeping a sound mind and a sound body.

So wrote Francis Bowen, offering what seems at first glance to be a simple recipe. But how simple is it really?

Who decides what "thoroughly good" means? Is being "good" a willingness to abide by the mores of our times? In past centuries that meant rigidly repressing sexual impulses far beyond what the average person could manage. It also meant condoning slavery. Throughout the centuries it has meant the legal subjugation of women. In our own day, full equality for everyone remains an elusive goal.

So-called "justice" in our land is still far from just. By a huge majority, white males still pass our laws, interpret them, and decide who will judge us when we flout their white-male determined rules. In addition, we not only still eat animals. We also abuse them, and at our own whim use them as experimental creatures, often subjecting them to intense pain or keeping them under appalling conditions. In this context, what does being "a thoroughly good man" mean?

Does it mean following what our parents taught us? Or our favorite teachers? What our minister preaches? What the Bible says? Or the Koran, the Talmud, the Upanishads, or the teachings of Buddha? Should we read Confucius, or Lao-tse, or the tarot cards, astrological charts, or tea leaves?

What if trying to follow what all these sages tell us leaves us repressing so much anger we are time bombs just waiting to explode?

What if trying to be "thoroughly good" to others leaves us so depleted we have nothing left with which to be good to ourselves? Surely our own needs count for something.

What about our own instincts, our own intuitions, the still little voice within? Are they a reliable guide? Possibly they are for many of us, or for most of us, but what if our interior voice tells us to go grab a submachine gun and kill off as many people as possible? Or to kidnap a three-year-old child in order to rape and kill her?

Where do we go for clear and easy guidance we can trust?

COMMENTARY

We can tune into our minds and bodies and listen.

If we are free of all pain, both mental/emotional and physical, the spirit and mind are in easy fit and all is well. If they fall out of fit, which causes disease, if we are listening we will hear.

The spirit will first try to inform the mind, but the mind being the frantic, flighty, experienced-at-dodging coward that it is, the body may pick up the message first and pass it along in the form of physical discomfort.

If we ignore this discomfort, popping pills to alleviate it, the spirit won't give up. It will only become more insistent. In time, if all else fails, pain will be the language with which the indwelling spirit finally manages to speak to us.

If we'd like to be well, we must listen.

We don't have to fear, as we first begin to listen, that we'll have any trouble hearing. If we ignore discomfort and it then progresses, the pain when it hits won't come through overly soft or muted so that we have to strain to hear. Rather, the "sound" of pain ordinarily comes through loud and clear, so much so that far from having to quiet down to hear it, we most often wish we could somehow, some way, shut it off. But surely we all know by now that it's dangerous to drug ourselves past a certain point to shut off pain, and at other times it's useless. The pain persists regardless of how many pills we take. So here's an alternative suggestion:

Learn the language of the body and listen to it, if you'd prefer to be pain free.

HYPOTHESIS:

To remain healthy, we need to have our conscious minds deal with and discharge negative emotions instead of shrugging them off onto the body.

Note: For some people, especially for those severely abused in childhood, the burden of negatives can be so heavy as to overwhelm the mind, to where it can't carry the burden without breaking or splitting. In such cases, being able to transfer part of the burden onto the physical body in the form of various problems may be a forward step, allowing healthier mental and emotional functioning.

For most of us, dealing with fairly mild negatives, this is not the case. Our physical problems are far more likely to be the result of laziness or cowardice. Rather than confront a loved one about what is really bothering us, or to devote the time and effort needed to dig deep inside to see what we're really made of, it's easier to suffer a cold, come down with stomach flu, or to have increasingly stiff joints. From there we can go on up the scale to serious physical breakdowns.

* * * * *

Isn't the above playing the same game that critics of Dr. O. Carl Simonton once accused him of playing, a rotten, "Blame the Patient" game?

Yes, it is, absolutely. It is placing upon the "victim" of any illness full blame...full responsibility...for any dis-ease she suffers. Underlying this is the belief that all of us are supremely creative beings who form our own bodies to suit ourselves, that we are in full charge of our own lives, creating them as we go along, and if we who occupy our bodies aren't responsible for their care and upkeep, who is?

The notion that human beings have free will and are responsible for their actions...including internal actions that avoid or bring on illness...is not a new idea. In 1926, Georg Groddeck published a book *Das Buch vom Es* (*The Book of the It*) which argued that all diseases have a psychogenic or psychological source. Groddeck believed that people created their own diseases,and there was no need to seek any other cause.

Just as people made themselves ill, it was within their power to make themselves well again, according to Groddeck.

Any such belief flies in the face of current social thought. As the 20th Century progressed, Americans were held personally responsible for fewer and fewer of their problems.

If children are restless and don't want to or can't sit still in school, it's not their fault. They are hyperactive or suffer from Attention Deficit Disorder and need to be drugged. If women fly off the handle due to premenstrual tension, this is caused by a recognized disability, for which there is a medical cure. Alcoholism doesn't result from weak character. It's a disease. Crimes are often fully or partially excused as due to extreme emotional distress. In every direction we look, we are understood, forgiven, not held responsible, let off the hook.

The move in this direction is heating up. Combine the completion of genetic coding with powerful new technologies and the brain will be under surveillance as never before. The belief of eager young neuroscientists is that they will be able to confirm current neuro-scientific theories about "the mind", "the self", "the soul", and "free will." All will be swept away, they predict, shown to be non-existent, exposed as nothing more than illusions. There not only is no "free will." There is no "I" to exert it.

Everything is predetermined. We do what we do because we have been hardwired to do it. We are composed solely of matter and water, strings of molecules connected to a computer known as the brain, which, while going about its normal operations, creates such illusions as the absurd notion that the self exists and has free will.

If this deterministic view becomes established belief, no human being will be held responsible for anything. If young males grow up to be criminals, it is not their fault. They were hardwired to behave that way. If others are raised in extreme poverty, in broken homes, amid violence, but manage through hard work and talent to become happy and successful adults, it is not to their credit. At birth they were hardwired to achieve this.

Our erroneous faith that we are a "self" is known as the "ghost in the machine" fallacy, the quaint belief that somewhere inside the brain there is a ghostly "someone" who runs things. The illusion of a "self" is created by our neurological systems. The human mind is totally dependent upon, and does not exist apart from, the physical brain that gives rise to it. So claim the eager young neuro-scientific determinists.

In opposition to this is the belief that our minds and our brains are different and distinct. Wilder Penfield, a pioneering Canadian brain surgeon, believed that the mind experienced life while all the brain did was record it. When he began his work in the 1930s, science had not yet decided that the mind was an illusion, a "ghost" created by our neurons. Experience of all kinds persuaded not only Penfield but other researchers that the current neurological view of mind is mistaken, and that our minds, with their strong sense of self, have remarkable powers that cannot be explained by the workings of our physical brains.

An additional proof of the mind's independence, and one that I think is all but impossible to refute, is the out-of-body experience, which has been known about and recorded throughout human history.

Anyone who has ever been "out-of-body" has experienced the ability of the mind to function with all its senses intact apart from its purported base of operations in the physical brain. Such journeys are said to keep intact the tie between mind and brain via the "silver cord," the invisible tethering device that allows the mind-spirit-soul to wander away while staying attached, but this attachment has been seen as necessary in order to keep the body alive, not the mind-spirit-soul. The mind stays attached so that it can once again inhabit and give life to its body.

Neuroscientists claim they can find no evidence of a "self" within the brain. In fact, three-dimensional electroencephalography shows that there is not even any one place in the brain where consciousness or self-consciousness can be located. But love...joy...peace...anger...fear...hate...pity...cannot be found there either. Does this mean all these emotions don't exist?

If I were diagnosed as having a serious, possibly fatal disease, here is what I would do.

First I would ask myself whether I wanted to be cured, if I wanted to survive it, and I would instantly reject an immediate, glib response, "Of course I do." I would think about it long and hard, meditate on it, try to dig deep within myself to the level of absolute certainty where I know I would find an honest answer.

As Dr. Simonton pointed out in his book *Getting Well Again,* ill health brings rewards. Friends offer comfort and support. We can stay home from work to rest up. The more serious the illness, the more likely these benefits will accrue. If we're diagnosed with a potentially fatal illness, once word gets around we may hear from people we haven't heard from in months or years. Our spouse becomes extra loving. Our children rally round. We are thrust into a spotlight that quite possibly we have never enjoyed before. This can become a powerful incentive not to recover too quickly.

Possibly deep down we don't wish to recover at all. If on some level we are tired of the struggle life is, tired of trying to succeed while meeting one defeat after another, tired of chasing after happiness or a peace of mind that keeps eluding us, we may find a fatal illness comforting. It is rarely condemned the way suicide is. It won't throw our loved ones into the turmoil that a self-inflicted gunshot to the head would. If deep inside we'd just as soon toss in the towel, a fatal illness is a respectable way to depart this life.

If after intense introspection, I decided I *did* want to beat my possibly fatal illness, I would memorize as a mantra the message: *The life that made me sick is not the life that will make me well.* I would print this out and post it where I couldn't help but see it...and ponder it...daily.

I would pay close attention to the area of my body most affected by my illness. I would read up on the normal functioning of that area and zero in on its symbolic meaning for me. I would then do my best to relate this meaning to my emotional attitudes and my behavior. In other words, I would try to figure out what message I am being given, in what way I need to change. Then I would dive into myself, however painful I found this, and try to make the change that's called for.

To interject here a pessimistic note: In physics there is a law formulated by Robert Hooke known as the *Law of Elasticity*. This law states that a force working on a stationary mass produces distortion in the mass. Remove the force and the mass will snap back into its original shape. If the mass does not resume its original shape, the force has distorted the mass beyond its elastic limit.

No substance is totally elastic. Stretch a rubber band and let go. It will snap back into its original size...almost but not fully. Stretch it hard enough, long enough, and it won't snap back at all. The force applied to it will have exceeded its elastic limit.

The law of elasticity may well apply to our bodies. Force applied too long and too hard to our working parts may distort them past the point where a release of pressure will allow them to snap back. The damage done may prove to be irreparable.

This does not mean that we cannot improve our health by releasing the distorting pressure. We can. But the degree of improvement may depend to an unhappy extent upon the amount of damage already done.

To return to the listing of what I would do:

I would set aside at least an hour each day for meditation. Numerous studies have shown that meditation confers great mental, emotional and physical benefits. These benefits are greater than those we reap from simply sitting quietly and relaxing, even though that too is good for us.

As I relaxed before getting deep into meditation, I would try to zero in on why I wanted to continue living and what the purpose of my life would be. More importantly, I would try to dive in deep enough to get in touch with all my stored up negatives and then I would do my best to discharge them one by one, all anger, resentment, envy, jealousy. I would face my possible death as serenely as possible, acknowledging that no one escapes this final journey, but then I would gently remind myself that I wasn't ready quite yet to make this transition. Someday…someday peacefully and serenely…but not quite yet.

I would also pray and ask others to pray for me.

I would watch movies I think are funny, or read clever books, or talk to people who amuse me, and laugh as much as possible…

If I was physically able to do so, I would continue doing or begin doing some kind of volunteer work, for there is nothing quite like helping others to make one feel good about one's self.

I would immediately alter and purify my diet. I would stop eating all meat (if I were a meat eater, which I'm not), and almost all prepared foods, substituting raw food to where it comprised 75% to 80% of my diet, supplemented by soy products and beans. In addition, I would check out diets in books and magazines at a health food store, and keep up on the latest nutritional studies relative to the illness I was fighting.

As far as more conventional treatment goes, I would listen carefully to whatever my doctor said, ask all the hard questions, demand to know the pros and cons of any proposed treatment, do some research on my own, and decide whether I would benefit from what was offered. If I felt I would, I would accept it as one additional tool in my quest for a cure. But never would I allow myself to sink into the lazy, comfortable belief that it was the doctor's job to cure me and that all I had to do was follow orders.

According to Dr. Bernie Siegel in his book, *Love, Medicine and Miracles,* "good" patients, those who follow doctor's orders so that their doctor can "cure" them, are the ones who rarely beat the odds. Rather they die right on schedule. If your wish is to die right on schedule, fine. Otherwise act as though your health matters to you and participate in the healing process.

COMMENTARY:

If we would stay fit, we can begin by looking upon life as a seminar on anger management. If we repress all angers, sweeping them under the rug, we are in trouble. If we act out angrily with our mouths, our fists, with knives or with guns, we are in trouble.

Somehow we must learn to practice a detached scrutiny of ourselves so we know when anger is there, or fear, the other face of anger, so we can figure out a way to deal safely with these negatives. If the mind deals with them, the body won't have to.

The indwelling spirit may seem elusive, hard to keep up with, difficult to communicate with, but if we listen to our bodies and pay attention to our thoughts, we can learn how to cleanse ourselves so that no unhealthy build-up of negatives occurs, allowing us to enjoy the manifold delights of bursting good health.

20

Verification

◆

To Test or Not to Test, Here Are Some Questions

○ ○
It is wise to listen, not to me but to the Word, and to confess that all things are one.

——Heraclitus

Throughout this book speculative comment has led to the formulation of hypotheses. For these hypotheses to be elevated to theory, they must be corroborated through observation and testing. If they withstand this test...if predictions made hold up and are not disproved...then in time, with more observation and testing, theory can rise to the lofty heights of accepted law.

All the hypotheses herein have been grounded in a view of man as a godlike creature totally in control of his body and his life, and hence responsible for both, from the thickness of his eyelashes to the location of his cancer. This view runs counter to the prevailing winds of our times, which do their best to make helpless victims of us all.

No doubt it is comforting, in a way, to view oneself as a victim. It's a neat, easy way to shrug off guilt. I'm not responsible for this mess I find myself in. My genes...or germs...or unfavorable external circumstances...or fate or God...something outside myself, not under my control...is doing this to me. No matter that such a view puts us in, and keeps us in, an enslaved condition.

Only as we accept full responsibility for every aspect of our lives do we have any hopes of being free. No one who can be victimized by forces external to himself can be considered free.

My contention has been throughout this book that we create and control our own bodies. Here are possible studies that could be made to reinforce or disprove this connection:

If we are not at the mercy of our genes but instead form our own bodies to suit our own needs, then one of the things we decide for ourselves is our coloring. A newborn baby, having chosen parents of a certain race, would color her skin accordingly. This still leaves hair and eye color. If we choose eye color for ourselves, there should be a correlation between the chosen color and personality type or choice of occupation.

Let's consider two fairly common eye colors in the white race: Blue and green.

All colors have numerous associations and connotations.

As the color of the sky, blue becomes the color of peace and heaven, of heavenly peace. But as few of us reach this heaven while stuck here on earth, blue becomes the color of the wild, unreachable beyond, the color of defeat and depression.

To be blue-eyed is, in slang terms, to be innocent and gullible, while to be blue-nosed is to be strongly puritanical. Another slang use of the color mocks this innocence and Puritanism. Blue comic material and blue jokes are lewd, lascivious, and vulgar, suggesting the obscene.

To be blue-eyed is to be at peace, or to thirst for peace, which makes one a visionary or an escapist. How would this translate into occupational choices?

Green is the color of hope, of spring, of renewal, but though the earth renews itself with bursting vigor each year, we humans often finds ourselves unable to do so and jealousy and envy set in.

To be green-eyed is to have an inner liveliness, restlessness, envy, jealousy, to feel a need to spring up and exert control. What occupational choices would this type of temperament find most congenial?

PREDICTION

If studies were done correlating eye color with occupation, a higher percentage of blue-eyed people than is found in the general population would be found in these occupations:

> *Artists of all types... writers, painters, sculptors, actors... except for musicians. In addition, journalists, physicists, physicians, except for surgeons and psychiatrists.*

A higher percentage of green-eyed people than is found in the general population would be found in these occupations:

> *Composers, orchestra leaders, surgeons, psychiatrists, dentists, politicians, and those who deal in illicit drugs.*

If we are not at the mercy of our genes, but instead build out own bodies to suit our own needs, there should be a correlation between body type and personality traits.

Let's start at the bottom, with our feet.

Our feet are a part of our physical selves. They are our direct connection to the earth.

PREDICTION

People who are strongly materialistic, who care about material possessions and about their position on the economic/status ladder, who revel in sex and are driven by sexual needs... a personality type typified by politicians and drug dealers...will be found to have large feet in relation to their height and general body build.

People who are less materialistic, basically indifferent to the wealth/display of their physical surroundings, who couldn't care less that their house is not as fancily decorated as their neighbors' home, who happily drive the same old car until it falls apart, people who are in no way driven by their sexual appetites, will have small, dainty feet in relation to their height and general body build.

* * * * *

With our hands we reach out to others and to the world, to grab or to give. It takes a larger hand to grab successfully with than to give with.

PREDICTION

Materialists will have larger hands in relation to general body size, non-materialists smaller hands in relation to general body size.

* * * * *

The waist divides our physical selves from our emotional selves.

PREDICTION

Those for whom love is grounded primarily in sexual passion will have negligible waistlines. Those for whom the deepest, most profound love is based on a sense of intimacy and loyalty...those for whom parental or brotherly or religious love is as deep and profound a feeling as sexual passion...will have far more pronounced waistlines, i.e. much smaller waists.

* * * * *

The female breast symbolizes maternal love, so a positive relationship should exist between breast size and the depth and force of unexpressed maternal love.

PREDICTION

Women who have little or no maternal feeling to express will have little or nothing with which to fill a bikini top, whereas women who crave children will be more amply endowed. However, once these women give birth and happily express their maternal feelings, they will find their breasts flattening and tending to sag. There is little inner excitement and anticipation left within to keep them high and firm.

Fat is known to accumulate in those areas of the body where normal functioning has been blocked. A compulsion to mother that has been blocked will create overly large breasts, the larger the breast the greater the blockage in expression.

As women pass through menopause and begin edging into old age, another factor sets in. The breast symbolizes new life, and few women who outlive their fertility go springing forth to seek out new life in other forms. It's easier to continue along in the rut one has already carved out, no matter how dull and boring or even miserable that rut is. To jump out of the familiar and go off to challenge the unknown takes more energy, more emotional get-up-and-go than most middle-aged women have. Fat accumulates where functioning is impaired, which means that older breasts are heavier breasts, for most women.

For those older women who either climb out of their ruts or get forcibly thrown out, who then create exciting new lives for themselves, this breast heaviness won't develop. The new life charging through them will keep their breasts young.

* * * * *

The penis is the primary male sexual organ. In the eyes of children, sexual intercourse is often seen as an attack on the female by the male.

Our slang shows clearly that this conception of sex as an attack remains with us long after we have left childhood behind.

PREDICTION

Penile size will vary as to how worried or indifferent a man feels about being armed with a dangerous weapon. For those who forswear use of the weapon against the woman they most love and fear…Mama…and through her by extension to all women, it does not matter how large their weapon is, for in the depths of their being they have unloaded it.

This means that natural born celibates and homosexuals can grow large penises. (An article in the Los Angeles Times, dated May 21, 2001, confirms that gay men have larger genitalia on average than straight men. As Mama and all her stand-ins are in no danger from them, they can grow and take pride in larger weapons. Note: This prediction regarding penile size was first made in print in 1983.) Men who are incurable chauvinists, who in their deepest hearts feel superior to women and have not the least worry about assaulting them, will also grow large penises.

If a man is not a natural born celibate, or a homosexual, or an incurable chauvinist, the more he worries about attacking women with this fearsome weapon he possesses, the smaller will be his penis. If he can't quite give up his right to attack, but is fearful of hurting anyone, to keep the size of his weapon small becomes the best compromise he knows how to make.

* * * * *

We look upon burdens as something we shoulder.

PREDICTION

If life feels very burdensome and we worry about our ability to carry our share of the load, we grow broad shoulders. If life seems more of a lark, an adventure, and we have no fear of holding up our end of things, we have no need of broad shoulders and grow narrower ones.

* * * * *

In the male body the penis serves as the organ of both urination and ejaculation, which may be simply a plumbing convenience, or it may...as suggested earlier in this book...have an underlying significance.

PREDICTION:

In those individuals subject to sexual pressures, for whom sexual activity remains viable, the frequency of urination will correlate inversely with the degree of sex drive and consequent sexual activity.

Those who perceive themselves as having a high sex drive and whose sexual activities bear this out will also have a talent for holding their urine, will not possess overly active kidneys.

Those who perceive themselves as having a low to middling sex drive, and whose sexual activities bear this out, will have healthy, but bothersome, overly active kidneys. One or two drinks of any liquid and they will be glancing around for the nearest restroom while their more sexually driven friends have no such concern.

To state it another way: Those who can easily control their sex drive have a difficult time controlling the annoying frequency with which they have to urinate.

Those who can easily control the frequency with which they urinate have a more difficult time controlling themselves sexually.

If a study of these two functions finds that the predicted relationship exists, the implications are obvious, not only with regard to kidney disease but also with regard to assisting the sexually driven to gain control.

* * * * *

Studies could easily be done to verify or disprove any of the suggested correlation between bodily features and personality types outlined above. The following predictions regarding how we experience certain diseases could also be subjected to testing to either refute or confirm them.

Let's start with the common cold, which we can experience in a variety of ways.

PREDICTION

Those who feel fortunate, who see themselves as generally blessed, will feel guilty as they get ready to come down with a cold and will suffer from a vexing sore throat.

Those who feel unfortunate, who see life as having treated them shabbily, will have only minor soreness in the throat, if any soreness at all, but as the cold gets underway will have noses that pour and bothersome, lingering coughs.

* * * * *

Vision Problems.

Dr. William Bates, in his extensive study of vision problems, found that those afflicted with myopia could not, even when carefully covering their eyes to shut out all light, see an unrelieved black. Once the eyes could relax sufficiently to view, when closed and covered, unrelieved black, the myopic condition was cured, according to Dr. Bates.

Personality studies of myopes found that among other distinguishing traits they scored low on proneness to guilt.

PREDICTION

Psychological testing will show that myopes are relatively guilt-free because they do not hold themselves responsible for the depravity of the human condition. They have sharply curtailed their vision to avoid seeing any connection between themselves and this depravity.

* * * * *

Breast and lung cancer.

PREDICTION

Studies would find a correlation between repressed maternal feelings or repressed lesbian feelings and breast cancer, also a correlation between sexual identify problems...a deep-seated rage over sexual stereotypes...and cancers of the lung.

* * * * *

One final topic: Weight.

Americans are currently suffering, in almost epidemic numbers, a condition that most doctors consider unhealthy. They carry too much weight. Diet pills and diet books sell by the millions, yet most of us are still bothered by an inability to keep our weight down to where we would like it. Why?

What is the etiology of this national malady?

We are an affluent nation, of course, with rich, fattening goodies readily available. But why, when we want to be slim and trim, do we constantly stuff our faces with calories just because they are there? Is there some underlying reason that in our hearts, no matter what we think, we really want to be fat?

Consider this: If you had an extremely important paper that you had to place on a desk, a paper that you absolutely *mustn't* lose, how would you guard it? Keep it from flying away, becoming misplaced? Chances are that you'd carefully weigh it down with something.

Imagine this: You are in the kitchen of a house that has an underground basement. The door to the basement opens into the kitchen where you are. You know that a dangerous maniac is prowling around in the basement below and that if this maniac climbed the stairs and came through the door, he would do you harm. What to do? Well, why not move the refrigerator over against the door? The weight of that should keep the door closed and the maniac out.

If we are repressing an enormous amount of anger/fear that we are afraid might climb the steps, push through the door and enter the conscious mind, one way to feel safer is to barricade the door with a great deal of weight.

If we are so frightened by our own negative feelings that we don't dare let them see the light of day, to ease our tension we can bury those feelings under layers of fat. That way they won't be able to get us. We might not like our weight, but it's still easier than worrying that the maniac prowling around in the basement below will burst through the door and wreak God knows what havoc. Anything is better than living with the terror of that.

PREDICTION

Weight loss can only be successful as negative emotions are faced, dealt with and discharged.

Those who sincerely want to lose weight might risk peeking down into the basement just long enough to lose some of their fear of their own highly charged negative feelings. As they gain the courage to let those feelings sneak little by little into the conscious mind, their need to reassure themselves with love…food…will decrease and their weight drop off. Anger is a monster, but a monster we all have living within us. Better to tame those feelings, if possible, and allow them to live upstairs than to chain them down in the basement and live in terror of their escape.

*****

What with our drug use, both legal and prescription, our over-work...as a nation we work ourselves into a state of constant fatigue...our over-eating, our hours spent watching banal TV fare, and our eagerness to blame someone or some thing outside ourselves for everything that we think, feel or do, we seem to be indulging in an orgy of self-avoidance. If only we could stop being afraid and take a chance on getting to know ourselves. This is surely the single greatest favor we could do ourselves.

* * * * *

In the opening chapter I attempted to show that our bodies tell us that our lives depend upon oneness/unity. This underlying oneness/unity of life has already been demonstrated, and is accepted as proved, by some scientists working with subatomic particles.

In his fascinating book, *The Tao of Physics*, Dr.Fritjof Capra writes that most of us in our everyday lives are unaware of this unity of all things: Hence we divide the world into separate objects. He suggests that this division is necessary and useful. However, it is not the fundamental truth about the universe we live in. Modern science has found that the underlying truth is the oneness of all things.

Physicists David Bohm and B. Hiley wrote that the classical notion that there are some basic elements that are the fundamental reality of the universe has been discarded. Now interconnectedness of the entire universe is accepted as the fundamental reality. Within this unbroken unity, some parts are relatively independent but remain in essence contingent forms within the greater wholeness.

We may well be one of the parts who can behave with relative independence, but how often do we think of ourselves in this fashion? How often do we see ourselves as contingent forms within an unbroken wholeness? How often do we run our lives with full awareness that the interconnectedness of all things is the fundamental reality of the universe?

Dr. Capra suggests that not being aware of the unity of all things, instead dividing the world into separate objects and events, has been useful to us and even necessary. Otherwise we might have difficulty coping with our everyday environment. But, lacking awareness of this underlying unity of all things, how well has man coped? Look at our history. Ever recurring wars, plagues, crimes, poverty, exploitation...this is coping? One wonders what the history of the world would have been had we not made this artificial division of the world into separate objects and in consequence *not* been able to cope.

> *The fundamental reality of the material universe, according to modern physics, is interconnectedness, inter-relatedness, interdependence, unbroken wholeness and unity.*

If life depends upon oneness/unity...if the reality of the universe is unbroken wholeness...then any assault on this oneness/unity becomes an assault against life.

If we see our neighbor as fundamentally separate from ourselves, and this distorted view of him justifies our right to strike him, rob him, or otherwise harm him, we are striking at our own life roots as well as at his. If our neighbor needs help that we refuse to give, we are depriving ourselves of nourishment quite as much as we are depriving him.

If we see another nation as fundamentally distinct and separate from our own, and the populace of that nation as more wrongheaded than we are or as somehow not quite as human, and this distorted perception justifies waging war, in warring against such a nation and its people we are warring against ourselves, for the fundamental reality of life is that all of life is one. We cannot slaughter our "enemies" without slaughtering an aspect of ourselves. We cannot injure another without injuring ourselves.

The terrorist...the child rapist...the serial killer...the mass murderer...the straight...the gay...the fat...the thin...the young...the old... *each and every one is a part of us.* Only as we begin to feel this unity and act in accordance with it...forswearing hate, anger and revenge...can we truly know peace of mind and enjoy genuine good health.

* * * * *

A profound connection...a profoundly meaningful connection...has been repeatedly postulated in this book between the spirit and the body/mind. It has been theorized that the indwelling spirit forms the body, then stays on the job to oversee its course through life, forcing renovation...discomfort or disease...when this is required.

Our knowledge of our world comes to us through our five senses that have proven to be limited and which routinely deceive us. We are equipped to see radiation only within a certain spectrum. Waves produce sounds above and below the vibratory speed that we are able to hear. Our fingers, touching numerous surfaces, tell us these surfaces are solid when in fact material objects are never "solid," but instead are composed of tiny bits of mass whirling at dizzying speeds through vast, empty spaces.

An inner voice tells each of us that he or she is a distinct entity, while our senses...limited and deceptive...tell us that the bodies we inhabit are solid and separate.

Modern science tells us that we are wrong on the second count. What if we are wrong on the first one also?

SUMMARY:

> *We are each unique, but unique as part of the whole. We come into life with godlike powers to create ourselves: Our bodies, our minds, our hearts, our lives.*
> *Each of us, a creative force, is in essence a spirit...and all spirit is one.*
> *Hail to thee, brave spirit.*
> *Hold to hope.*
> *Flow with ease.*
> *Creative with love.*

0-595-24807-1